Passport to Past Lives

Passport to Past Lives

◆

The Evidence

Robert T. James, J.D.

iUniverse, Inc.
New York Lincoln Shanghai

Passport to Past Lives
The Evidence

iUniverse, Inc.

For information address:
iUniverse, Inc.
2021 Pine Lake Road, Suite 100
Lincoln, NE 68512
www.iuniverse.com

Transcripts of dialogues with the subjects while in hypnosis have been edited to avoid repetition, and the names of the subjects have been changed to protect their identities. All the other parts of the dialogues are exactly as appear in the recorded sessions.

ISBN: 0-595-31022-2

Printed in the United States of America

For My Subjects,

whose search for what it means to be fully human made this research possible.

What a great bunch of people.

Contents

Introduction

What are we to make of the almost universal human experience, that regardless of our religious or other beliefs, while in hypnosis, we humans seem to have memories of our own past lives?

Are these memories, figments of our imagination, some form of genetic memories inherited from our ancestors, cryptomnesia (information learned from another source and forgotten), or plain fraud and deception? Are they true windows into our own immortality? Because of certain circumstances, I found myself pondering these questions a few years ago.

I had no committed beliefs about rebirth, past lives, or religious doctrines such as reincarnation. I do not accept the premise that the existence of past lives should be accepted on trust or faith, or be based on books or texts that many may hold sacred.

In saying this, I am not in any way trying to ridicule or denigrate those who do accept spiritual matters on faith. I like to think that as a skeptic, I have an open mind on all spiritual doctrines; but accepting faith as a criterion in the search for truth is just not what I do.

I am not a mystic nor a medium. I can not close my eyes and contact people who have died. I have not had a vision on the road to anywhere. I am a skeptic; but a skeptic with an active and open mind.

I am also a pragmatist. I do research, examine the resulting evidence of my research and the credible research of others, and try to draw reasonable and rational conclusions from that evidence.

With this book, I am not necessarily trying to convince the reader of anything, except that something seems to be occurring that cannot be explained by established, conventional, psychological concepts. And, that whatever is going on is important.

I started studying hypnosis many years ago while an undergraduate in college. After obtaining my Bachelor of Science degree, I went on to obtain a Juris Doctor degree and practiced law for thirty-five years. My interest in hypnosis, however, never waned. Over the years I continued my studies, attending a number of schools that taught hypnosis, hypnotherapy, and forensic hypnosis, including supervised training in these skills. In 1986, after retiring from my law practice, I returned to college and for the next three and one-half years studied traditional psychology and psychotherapy.

In the early 1980s, a friend asked me to regress her back beyond birth to see if she could recall a past life. I was reluctant, but I did it. She went into hypnosis easily, and did indeed seemingly recall being a married woman, with a husband and a child, living in England in the 1800s. The experience sparked my curiosity, but didn't spur me into seriously pursuing the phenomena. It seemed a little "far-out" for serious inquiry.

Two years later, the same friend again asked me to regress her beyond birth. Again I did. She regressed to what appeared to be the same life as before. In answer to my questions, she described the same life in much greater detail. In that lifetime she recited with some emotion, dying by being run-over by a horse and carriage in the streets of London while still a young woman. I had recorded both sessions. In comparing them, I was very impressed with the consistency of the detail that she reported from session to session, even though they were two years apart.

Now I was more than just curious. As an experienced trial attorney, I found it highly unlikely that she would be able to remember these same details years later if this were a made-up or fantasized story. Knowing the lady fairly well, I was also convinced that she had not deliberately falsified the regression to further some agenda of her own. This experience caused me to decide to explore the past-life phenomenon in more depth.

Because of both my skepticism and my experience and training in hypnosis, I felt, (right or wrong), that I was a suitable candidate to investigate this phenomenon. I had nothing to prove or disprove.

Many people hold hypnosis in disrepute, partly, I suspect, due to the use of it for entertainment purposes by stage hypnotists, and by its bizarre portrayal in motion pictures and television. For many people, combining hypnosis with past-life research pushes the entire matter to the edge of credulity. If I had any doubt about this, my colleagues reminded me of it from time-to-time. ("Bob, take up golf. Leave that stuff alone;" "Sounds weird to me;" and so on.)

Before advertising for subjects and starting hypnotic sessions, I conducted a search to determine what research, if any, of the past-life phenomenon had already been done. Recent books and journal articles contained a great deal of pertinent information, although many of them ardently attempted to prove or disprove the case for past lives. Very few have tried to evaluate the evidence supporting the reality of past lives critically.

Only a few authors that I encountered at that time, Ian Stevenson[1], Helen Wambach[2], Linda Tarazi[3], and Rick Brown[4] seem to have conducted reliable basic research into the subject following scientific acceptable methods that could be verified and replicated. I refer to the published cases by these authors as the "hard Core" cases, that I examine in detail in Chapter twelve.

Other popular books and reports contain cases that are certainly suggestive that we have lived previous lifetimes and that some form of our individuality survives our physical death. To a skeptic such as myself however, all these cases and reports raised questions for which I could find no satisfactory answers. For example, a number of the subjects involved persons seeking assistance with mental or emotional problems. Is the experiencing of past lives while in hypnosis the normal capability of a healthy mind? Or only an artifact of therapy?

In some of the reported cases, the authors seemed convinced of the reality of past lives. Is the subject, knowing the belief of the hypnotist, merely responding to the relationship and trying to please the hypnotist? Would the results of the regression be the same were the hypnotist a skeptic, neither a believer nor a disbeliever, and the subjects were so advised?

Would the religious beliefs, and the extent of the involvement of the subject in religious activities of one kind or another, affect the results of a hypnotic regression to a past life?

How do sex, age, and/or the education of the subjects influence past-life regression?

Would the subject's expectations affect the results of the hypnotic regression?

Would the past-life regressions tend to confirm or rebut the experiences of those reporting near-death experiences?

In the event the evidence does indeed indicate that some form of our individuality survives our physical death, has lived in the past, and will probably live again, such evidence would contradict the current view held by many Christians and others in our culture, that such surviving entity was newly created by some divine source at the time of the physical birth or conception of members of our species.

To attempt to seek answers to my questions arising from the research of others, I conducted two different research projects into the phenomena of past lives. I structured my research in accordance with scientific methods for conducting psychological research.

My first research project was designed to just generally inquire into the past-life phenomena to see if normal, healthy adults could and would regress beyond birth. To begin, I ran advertisements in magazines of general, but not prominent, distribution in the Colorado Springs, Colorado, area. My intention was to proceed inconspicuously, but still attract mature, emotionally-stable subjects.

Fortunately, I was successful in attracting an extremely interesting, sincere, and emotionally stable group of subjects, who sought no personal gain by participating in this project.

In my first research project, I worked individually with 107 subjects, three of whom did not go into hypnosis. Of the 104 who did go into hypnosis, eighty-one reported memories of what seemed to be past lives.

In their encountered past lives, the subjects were frequently of a different sex than they are in their present lives, and were frequently of different races.

In PART ONE of this book, I give a brief Overview of Hypnosis and Past-Life Regressions, in order to show how past-life regressions came about.

PART TWO, in Chapters two through eleven, I explore many of the experiences my subjects encountered in my First Research Project in what seemed to be previous lives, and also I explore their experiences in-between lifetimes. My subjects were all healthy adults, of different ages, of both sexes, with different religious beliefs and philosophies, almost all of whom did not know each other.

In those ten Chapters, I have set forth a wide range of my subjects' past-life experiences in the belief that this would give all of us a better understanding of the past-life phenomena, and aid us in analyzing the "hard-core" cases published by other researchers as I do in Chapter twelve.

Because of the general nature of my First Research Project, no specific attempt was made to obtain verifiable data, and very little data was obtained that could be verified using readily-available resources. The research (as with most) gave credence to some propositions, but also gave rise to a great many questions. The most important questions occurring to me at the conclusion of my First Research Project, was whether verifiable past-life data could readily be obtained from the subjects, and secondly, if the evidence indicated that we did indeed survive our physical deaths, what was the origin and nature of that surviving entity? How does such phenomena fit into our evolutionary past?

These questions and a few others gave rise to my Second Research Project. I again advertised for subjects. Seventy-three persons responded. I worked individually with fifty subjects. One did not go into hypnosis. Of the forty-nine who did, forty-four regressed to what seemed to be lives that they had lived before. Again, my subjects were all healthy adults, of different ages, different sexes, different religions and philosophies who did not know each other.

One of my two primary objectives in this Second Research Project was to regress the subjects back to lives just previous to their present ones, to see if verifiable data *could readily be obtained.* I emphasize *readily be obtained,* because in several cases from sources other than from my own research, evidence of past lives as recalled by subjects have indeed been authenticated. However, most of these verified cases didn't arise by plan, but were randomly encountered, like in the Tarazi and Brown cases.

Secondarily, I planned to take the subjects back to the first time they lived on Earth *in any form.* In addition, I also planned to take the subjects into their mothers' wombs, just prior to their present lifetimes. Could we determine when the subject entered the fetus? Were they aware of their mother's feelings and emotions while there? Lastly, as "far out" as it may seem, had the subjects ever lived on another planet? If they reported that they had, I wanted to explore those experiences.

PART THREE investigates with you my subjects' experiences in my Second Research Project.

These experiences may very well not fit into the reality with which you are familiar. I'll confess, it was vastly different from my own.

I offer the material in this book, not to convince you of any particular philosophy or point of view, but merely to present the results of my own research, along with a discussion of what seems to be a few well-authenticated cases of other researchers. The final determination of the importance of this material and its relevance to your life is, of course, always with you.

PART I

Hypnosis and Past-life Regressions
—An Overview

We shall not cease from exploration
And the end of all our exploring
Will be to arrive where we started
And know the place for the first time.

—T. S. Eliot

1

Why do People Regress?

In the chapters following, you will encounter the experiences of a number of the past life experiences of other investigators, along with the experiences of the past-life experiences of my subjects. These subjects were of different ages and gender and with different religious backgrounds. Most, but not all, while in hypnosis, regressed to what seem to be lives they have lived before. Most of them are probably just like you; in that, if you were regressed you would most probably experience lives you have lived before.

Past-life experiences sometime arise spontaneously in peoples' minds and memories; but if memories of past lives actually exist, for most of us, they must be buried deep within our minds. If they weren't, we would be overwhelmed with the sheer magnitude of many past events ever-present in our "normal awareness." Unless we're dealing with the very rare occasion where a person consciously remembers having lived before, to access these past memories we must enter into some altered state of consciousness, by-passing our normal awareness, allowing such access.

An altered state of consciousness, achieved with the help of hypnosis, is the method I used in my research to access these long-buried memories of lives we have lived before.

Along with many of the individual experiences explored herein, you will encounter my discussions of different hypnotic and past-life phenomena. The subjective nature of hypnosis and the importance of trance depth in accessing the subconscious in hypnotic regressions are not readily understood by many, particularly by critics of hypnotic regressions. It was my belief that a discussion of these phenomena as such phenomena arose in the regressions, was the best way to illustrate and explain what was happening.

However, to put my subjects' experiences in proper perspective, I believe that it may be helpful to briefly review how the exploration of past lives with the use of hypnosis came about.

3

Ancient times are replete with references to hypnotic-type practices. The Egyptians and the Greeks had rituals that lead the participants into trance-like states. But the more modern hypnotic related activities really began in the 18th century in France with a physician named Franz Anton Mesmer.

Mesmer developed a system of healing that he called Animal Magnetism. He believed that invisible fluids in the body that could be affected and controlled by magnets caused the trance-like states that occurred in his subjects. This theory was soon largely discounted although the practitioners that followed Mesmer using trance-like states with their patients were generally known as practitioners of Mesmerism.

A form of Mesmerism was used successfully in the 19th century as an anesthetic in major surgery by two English physicians, John Elliotson and James Esdaile.

James Braid, a physician practicing in England at that time (who coined the term "hypnotism"), was a supporter of the concept that the basis of Mesmerism was psychological, and not a physiological condition. He devised an induction system using eye fixation on a bright object, followed by sleep suggestions. He is generally considered the father of modern hypnosis.

Other medical practitioners of that period struggled to determine if Mesmerism was physical or psychological in nature. In the late 1800s, a famous controversy broke out between two rival schools of thought in France. Dr. Jean-Martin Charcot of the Salpetriere School in Paris thought that hypnosis had its basis in a nervous disorder, and Dr. Hippolyte Bernheim of the Nancy School, maintained that hypnosis was a psychological phenomenon created by the subject's reaction to suggestion.

Modern hypnotists don't generally consider themselves followers of either of those schools, but would probably acknowledge Bernheim's concepts as the more relevant of the two approaches.

From 1887 to 1889 Freud used hypnosis in his psychoanalytic practice to explore the subconscious minds of his patients. But he then abandoned it in favor of just having his patients relate whatever came into their thoughts; a technique called "free association."

Interest in hypnosis in the early 1900s generally declined until psychiatrist Milton Erickson's use of altered states of consciousness in psychotherapy became well known in the 1930s. Erickson is probably the best-known American practitioner of hypnotherapy in the 20th century. He died in 1980 but has a strong following today among hypnotists and hypnotherapists.

Because World War II presented so many opportunities for the use of hypnosis in the treatment of war-related mental and emotional disturbances, hypnosis again became a subject of general interest in the mental health field. Numerous technical journals and societies dedicated to the study and use of hypnosis came into existence and flourish today. In 1958, even The American Medical Association acknowledged "the validity of the various phenomena elicited by the hypnotic techniques."

The mind-mechanisms underlying the nature of hypnotic trance-states are still debated with no compelling consensus; but that such a trance state exists is no longer questioned by knowledgeable investigators

In modern times, "age regression", that is, regressing a patient back to childhood with hypnosis to explore the basis of emotional disturbances in lieu of free association, is regularly performed by many mental health practitioners.

In the 1950s and '60s, there was extensive research into the body reactions of subjects when age-regressed to young ages, to determine if such reactions were real. Studies demonstrated that in child-like states, the subjects had bodily reactions appropriate with the age to which they had regressed; sucking, eye movements, toe extensions when the sole of the foot was stroked, (Babinski Reflex), etc. Intelligence and Rorschach tests were given to subjects in their regressed states, changes in handwriting were tested, and other similar tests were performed. Accurate recall of child-hood memories was able to be verified.[6]

Also in the 1950s, individual practitioners began experimenting with their clients, age-regressing them back before birth in an attempt to access the basis of trauma arising in past lives that could be responsible for present-life emotional disturbances. Regressing a subject back beyond birth as we do in past-life regressions, is merely an extension of the technique of age regression.

"Past-Life Therapy" as a school of therapy thereafter developed and is alive and well today. Those interested in this type of therapy have their own organization, their own regular newsletter and a journal. Similar organizations now exist in the Netherlands and in Brazil.[7]

As a student of hypnosis for many years, I first learned the then commonly used induction method of directing the subject's attention to a bright object, followed by monotonous, repetitive instructions of physical relaxation, drowsiness, mental relaxation, describing the signs of approaching hypnosis, and eventually taking the subject into a hypnotic-state (shades of Braid). The process usually took from five to twenty minutes depending upon the subject. Looking back, I often wonder if rather than leading the subjects into hypnosis, we weren't just boring them into an altered state of consciousness.

In recent years, the traditional progressive suggestion induction method has been refined to accomplish an induction in a minimum of time. There are now many rapid induction techniques in use. Skilled hypnotists generally find some particular method that works well for them and with which they are comfortable. An example of the rapid induction method I generally use is illustrated in Chapter Two with a lady named Carol. It usually takes from about three to four minutes to put a willing subject into a reasonably deep state of hypnosis using modern rapid induction methods.

Several interesting books have been published in the last few years recounting therapists' experiences with individual clients who encountered past-life memories. I have made reference to several of these books in the following chapters, and in the "Notes" that follows at the end.

Now, to answer the question "Why do People Regress?" that I asked at the beginning of this chapter. As demonstrated by my research, the ability to regress back before birth while in hypnosis to what seems to be lives they have lived before is a common, natural function of most normal, healthy, persons.

Since most (but not all) of my subjects in both of my research projects did regress back beyond birth, a more pertinent question is, "Why do some people not regress?"

I can speculate. Some people are reluctant to give up control and let the hypnotist lead them back. Or a subject's subconscious mind may be telling the subject "Don't go back. There are things there you can't handle and don't want to know." Others may not have a past life to go back to.

As my research indicates, the depth in the hypnotic trance that a person achieves does have an effect on whether a person regresses back before birth. So the subject who doesn't regress may not have been sufficiently deep in a hypnotic trance-state to access their past-life memories.

Although it is normal to regress while in hypnosis, as you can see from my speculation, truthfully, I really don't know with certainty why some people don't regress.

To explore the past-life experiences of those who were able to regress back before birth with or without the use of hypnosis is a journey into terrain that may be new and foreign to you. If you are a skeptic, just curious, a serious student of the possibility that we humans have lived before, or already a believer, this journey is open to you. The only requirement is an open mind, and a willingness to temporarily suspend any preexisting beliefs that would place limits on your ability to expand your concept of reality.

PART II
The First Research Project

—Recalling past lives in normal, healthy adults.

We have found a strange footprint on the shores of the unknown.
We have devised profound theories, one after another,
to account for its origin.
At last we have succeeded in reconstructing the creature that made the footprint.
And Lo! It is our own!

—Sr. Arthur Eddington

2

The Enemy is Coming

I was very pleasantly surprised, and a little overwhelmed, when 166 people responded to the advertisements for my first project. I sent each respondent a letter explaining what I was doing, information concerning many common misconceptions about hypnosis, and an Information Survey. (A copy of the Information Survey is set forth in full in Appendix A)

One of the first to responded was Carol. Her Survey disclosed that Carol was married with children, in her late 40's, white, well-educated, Protestant, and moderately involved in religious activities. She believed in some form of life after death, and she was uncertain as to her belief in past lives or rebirth.

Working with Carol was a joy. She was friendly, articulate, and self-confident. She easily went into a medium-deep state of hypnosis.

In inducing hypnosis with Carol I used a modified Dave Elman rapid-induction method.[8] Carol's eyes followed my fingers, and as I moved my fingers down across her face, she closed her eyes. I had her concentrate on relaxing the muscles around her eyes to the point where she was so relaxed, her eyes wouldn't open without stopping the relaxation. I then raised each of her hands a little way, one after the other. I dropped her hands and had her feel a wave of relaxation flowing from the top of her head, down through her body to the bottoms of her feet. This generally produced the necessary physical relaxation in a subject.

I then had her count slowly from 100 down, saying the words "deeper asleep" with each number. I told her that as she got down to about ninety-six to ninety-five, the numbers would just fade from her mind, and she would fall deeper and deeper into hypnosis.

Next I had her see herself on the tenth floor of a building, standing in front of the elevators. The doors opened, she got in. And as she traveled down to the first floor, I further deepened the hypnotic trance. When she reached the first floor, the elevator doors opened. Before her was a large, white cloud that would take her traveling back into time, back before she was born into this lifetime. "It's a safe, friendly cloud. You won't sink through or fall off."

I used a self-reporting system to help in determining the depth of the hypnotic trance.[9] (0 being awake—ten deep). The importance of trance depth in regression is that if the subject is in a light to medium trance they often are analyzing and at times censoring the memories they encounter, lessening the likelihood that we are accessing true past-life memories. Carol reported being very deep and I actually had to lighten her trance so that she could speak clearly.

I suggested to Carol that when she saw herself in another lifetime as she traveled back on her cloud, the cloud would stop and she would signal me by raising her right index finger. After she had signaled me that she was present in another lifetime, the following transpired:

RTJ:	Now without any fear, pain, or discomfort, come down from your cloud.
	Tell me now, how old are you?
C:	thirteen.
RTJ:	Now look down at your arms. What color is your skin?
C:	It's brown.
RTJ:	Are you a girl or are you a boy?
C:	I'm a girl.
RTJ:	You're a girl. What's your name?
C:	(No response.)
RTJ:	There's someone standing right next to you and they're talking to you.
	They're calling you by name. What name do they call you?
C:	Maria.
RTJ:	All right Maria. Now tell me, are your mother and father still alive?
C:	Yes.
RTJ:	What's your father's name?
C:	Tarento (phonetic).

RTJ:	What's your mother's name?
C:	Isabelle.
RTJ:	Now I want you to go forward in time. Go forward in time until you're about twenty years old. Look around you and tell me, where are you now?
C:	In the market place.
RTJ:	Are you buying something in the market place?
C:	I'm walking.
RTJ:	You're walking. All right, look around and tell me, what are they selling in this market place?
C:	All kinds of food…and material.
RTJ:	Are you buying anything? Are you buying any kinds of food or material?
C:	No. I'm hurrying.
RTJ:	Why are you hurrying Maria? Where are you going?
C:	To the shop.
RTJ:	Why are you going to that shop?
C:	A white shop…'cause the men are all waiting there.
RTJ:	All right. What are you going to tell them when you get to that shop?
C:	(Here Carol is speaking with some urgency.) A big secret. That the enemy is coming.
RTJ:	That the enemy is coming?
C:	Yes. Uh-huh.
RTJ:	Who is the enemy?
C:	They're on horses.
RTJ:	What country is this in Maria?
C:	The walls are all white. It's hot. The streets are all cobblestone.

RTJ:	Look around. Are there any animals that you can see?
C:	Dogs.
RTJ:	Are people traveling using animals?
C:	In carts.
RTJ:	All right. Now we're getting closer and closer to this shop. And it's getting clearer and clearer now what country this is. What country is this?
C:	It's Italy, I think.
RTJ:	What year is this, Maria?
C:	It's 18…1864.
RTJ:	Now we're getting closer and closer to the shop. Tell me Maria, do you have a mate?
C:	No.
RTJ:	No. All right. Now you're right in the shop where the men are and you're telling them the enemy is coming. How did you find out?
C:	I went to their camp.
RTJ:	When did you go to their camp?
C:	Last night.
RTJ:	Did you hear them talking? How did you know they were going to come?
C:	They don't pay no 'tention to me. I'm a woman.
RTJ:	Are these men that are coming…are they Italians too? Who are they?
C:	They've been taking over everything. And they're coming to take our town now. I just got out.
RTJ:	Look down at yourself. Tell me, what are you wearing Maria? You're in the shop, talking with the men.
C:	It's soft…and ah…its got a head piece on it…and I've got a lot of pretty jewelry on.

RTJ: Tell me about this jewelry. Look at it. Is it around your arms or neck?

C: Both. I like jewelry.

RTJ: What does the jewelry look like?

C: It's got beautiful colors…and stones…I think it's silver.

RTJ: Where did you get this jewelry from?

C: I get it from men.

Here I was having a little fun with Carol.

RTJ: What do they give you this jewelry for?

C: Hmm. (very coyly). For being their friend.

RTJ: Now look down at your feet. Do you have anything on your feet?

C: Sandals.

RTJ: Now go forward to the next significant event. You just told the men that the enemy is coming. Tell me, what's going on now?

C: They're fighting.

RTJ: What kind of weapons do they have?

C: Swords…and…they use anything they can get their hands on…clubs.anything. (With emotion.) They thought we didn't have anything. But we did.

RTJ: Why are the men attacking this place? What do they want?

C: They just loot.

RTJ: Go forward to the next significant event. What's happening now?

C: It's quiet. The war's a long time ago.

RTJ: How old are you now, Maria?

C: I'm sixty-seven.

RTJ: Do you have any children, Maria?

C: Yes…but it's not very clear.

RTJ: All right, now without any pain or fear, I want you to go forward to the day that you died in that lifetime. How do you feel now? You've just died in that lifetime.

C: Satisfied.

RTJ: Look back at that lifetime, now. What was the purpose? What purpose did you have in that lifetime?

C: I was foolish. To get the information necessary. They thought I was a stupid kind of girl. I wasn't stupid. I know…I got the information they needed. And we're all okay. Now they don't think I was a harlot. Now they know. I'm glad they know. I wasn't dumb. A harlot or something.

RTJ: Now you just died in that lifetime. Tell me, can you see your body?

C: Yeah.

RTJ: Where are you that you can see your body?

I want to leave Carol at this point in her regression. In Chapter Five we'll explore the remarkable in-between lifetime experience Carol had after dying in the above lifetime, together with the experiences that many of my subjects had after they died in past lives.

The above transcription gives you an idea of what a past-life experience is like, although the facial expressions, emotions such as fear or anger, inflections, nuances, timing, pauses, and the like exhibited by the subjects are lost in just a transcript. The intensity of the expressions do indicate that actual events are occurring in the subject's past life. Frequently the subjects radically change their manner of speaking while in the trance personality. As I mentioned, Carol is a well-educated lady with graduate degrees. She is well spoken, and is used to public speaking, and speaking before professional groups.

In her personality as Maria, her manner of speaking and her use of words changed dramatically. She used abbreviated speech in some instances, "'cause the men…", and "They don't pay no 'tension.". The last sentence also contains a double negative, which Carol would not use in her normal speaking manner.

Notice also, her rambling, disjointed answer to my query, "What was the purpose", and her sensitivity to her life-style as a lady of the streets.

This change of personality while in the trance adds to the depth of the mystery in trying to evaluate the past-life phenomenon. It's very seductive in that it goes way beyond just the rote telling of a story. Fantasy? Role Playing? Some form of psychological projection?

Let's pursue these matters further and meet Jane, who regressed to being a pygmy, living in the jungle.

3

The Canoe Turned Over

I was not only overwhelmed by the number of people who responded to my advertisements, but also surprised by the composition of the respondents who eventually participated in my research.

Those who elected to participate in my first research ranged in ages from twenty-one to sixty-three, with the highest incidence occurring in the ages from twenty-nine to forty-nine. Seventy-eight were female and twenty-six were male. Eighty-percent had some college, ranging from one to nine years. Three had Masters degrees and two had Ph.Ds.[10]

One of my early participants was Jane, who arrived at the session with an interested friend. The friend was placed some distance away from our hypnotic session, where our activities could be seen, but not heard.

Jane is a single lady in her mid-forties, white, college graduate, with no prior experience with hypnosis. Her Survey indicated that she had been raised in a traditional Protestant religion in middle America, but she marked "other" to my question concerning religious preference. She indicated her degree of involvement in religious activity was "deep". She also indicated that she believed in some form of life after death, and that she believed in reincarnation. Fifty-seven percent of my subjects indicated they believed in some form of rebirth after death.

As with all my subjects, I first had her stand, arms out-stretched, with eyes closed. I took her through a preliminary suggestibility test (arms rising and falling). I have found a strong relationship between good responses on that test and the ability to quickly achieve a medium-deep trance state. The response from this test also gives me some idea as to the extent of the deepening techniques needed to obtain the most effective state of hypnosis.[11]

Jane's response to the test was Excellent, and her self-report hypnotic depth was six. She quickly went into a medium-deep state of hypnosis. As with all the subjects, I put her on her traveling cloud and then regressed her back to when she was much younger; first to age fifteen, and then to age five. In both instances I brought her down from her cloud and had her briefly described where she was, who was with her, and what she was wearing. At both of these ages, I instructed her that she would only see pleasant scenes. I was not searching for dysfunctional childhood experiences in this research.

Regressing the subjects to these two present life time-periods first were helpful stepping-stones for them to go back to a past life. The subjects' visits to their childhoods were pleasant. And also acted to verify the regression as they experienced it.

Jane was now very relaxed, eyes closed, and speaking in a very soft voice. I repeated her answers to make certain that they would be recorded on the tape.

After first visiting her two younger ages in her present life, the following transpired:

RTJ: Now, let's go back up on our cloud. Now we're going back...back into time and if you should experience any pain or discomfort, or if you should become afraid. Any time you want to, you can go right back up on your cloud, where it's safe and secure. Now the cloud is going back into time, back...back. Jane, the cloud is going back to a time before you were born into this lifetime. You're going to be able to speak to me in English or any other language you might encounter. You're going to be able to look around and tell me what you see and tell me what you feel...without waking up. And when you see yourself in another lifetime I want you to signal me by raising your right index finger, that I'm touching gently now. As you go back, you will continue to listen to my voice...and each word that I speak will put you deeper and deeper into hypnosis.

After a short time, Jane signals by raising her right index finger.

RTJ: Oh that's fine. Now come down from your cloud. All right, now tell me, how old are you?

J: fifty-four.

Notice that the age to which she regressed was older than Jane is now.

RTJ: You're fifty-four. And are you a man or are you a woman?

J: I'm a woman.

RTJ: Now I want you to look around. Look around and tell me, what do you see?

J: I see my people.

RTJ: You see your people. Now tell me about your people. Who are they? Who are your people?

J: We are the people…of the forest.

RTJ: And tell me, what's your name?

J: Inkwa. (Phonetic.)

RTJ: Now Inkwa, I want you to tell me what you see. Tell me where you are?

J: I'm outside…in the middle of the village.

RTJ: Does the village have a name? Tell me?

J: No.

RTJ: It doesn't have a name?

J: It's just our home.

RTJ: And are your mother and father still there?

J: No. They're dead.

RTJ: What was your mother's name?

J: Kawa. (Phonetic.)

RTJ: And what was your father's name?

J: Inkgwa. (Phonetic.)

RTJ: And do you have a mate?

J: No.

RTJ: Have you ever had a mate?

J: No.

RTJ: You never had a husband?

J:	No.
RTJ:	Do you have brothers or sisters?
J:	No.
RTJ:	Do you have any children?
J:	Yes.
RTJ:	And how many children do you have?
J:	I have three.
RTJ:	Tell me, what's their names?
J:	Bawa...(I'm unable to phonetically spell the second name), and, Coseta.
RTJ:	That's fine. Now I want you to see yourself, sitting down and eating a meal. Tell me, what are you eating?
J:	Fruit.
RTJ:	Fruit?
J:	Grain. Small grain and fruit.
RTJ:	Small grain and fruit.
J:	And a paste, made of root...white.
RTJ:	Now I want you to look down at your self. Are you wearing any clothes?
J:	No.
RTJ:	Are you wearing anything on your feet?
J:	We make protection of vines and leaves.
RTJ:	I see. Now I want you to look at your arms. Look around your neck. Tell me, are you wearing any kind of jewelry?
J:	I have...a necklace...bones...bird bones.
RTJ:	Now look at your arms. What color is your skin?
J:	Black.

RTJ:	It's black. Now I want you to go forward. Go forward in that lifetime…'till the time you died. You'll be able to look at this without any pain or discomfort…and you'll be able to tell me what's happening. Go forward to the day that you died in this lifetime. Tell me now, how old are you?
J:	fifty-six.
RTJ:	And tell me…look around.
J:	I'm very old.
RTJ:	You're very old?
J:	For my people.
RTJ:	Look around and tell me, where are you now? Where are you on the day you died?
J:	On the river.
RTJ:	Is there anyone else there?
J:	Eight.
RTJ:	There's eight people there?
J:	Two canoes.
RTJ:	Two canoes. And tell me, why are you dying?
J:	The canoe turned over.
RTJ:	The canoe turned over?
J:	In the river.
RTJ:	Now I want you to pass right through that death experience.
J:	(Deep sigh.)
RTJ:	Now it's after you died. Can you still see your body?
J:	It's lost.

After awakening, in the post-hypnotic interview, Jane advised me that she was not only of the black race while in the jungle, she was also a pygmy. This points up a common occurrence that happened with almost all of my subjects. While within what seems to be a past life, they had experiences and saw things about which I wouldn't be aware enough to even ask. They would frequently describe these events in the post-hypnotic interview. I furnished tapes of the sessions to my subjects if they so requested (with the induction omitted), and several reported that when they replayed the tapes, they would almost relive the past-life experience, and would see things that they hadn't mentioned to me before.

I would also like for you to meet Kay and look at one of her past lives while a Christian in early Egypt and Rome. Kay is a white, married lady, with three children. In her mid forties, she is presently a graduate student. She stated her religion to be Catholic, and her degree of involvement in religious activities was "moderate." She indicated that she believed in some form of life after death, and that she also believed in rebirth after death.

Kay had an Excellent response to my hands rising-and-falling susceptibility test. Her self-report depth in hypnosis was seven. She readily went into a medium-deep state of hypnosis.

Kay regressed to being a woman, twenty-three years old, with light-brown skin, wearing a robe and sandals. She stated her name was Ruth Wile.

RTJ: Look around and tell me, where are you? You're twenty-three years old.

K: In a boat.

RTJ: Are there other people there with you?

K: Yes.

RTJ: Where did the boat start from?

K: Shore.

Note here, the exact, literal answer to my query. This frequently happens in hypnosis when the subject is in a deep state. If you ask a subject who is in deep hypnosis "Will you tell me your name?" Instead of stating his or her name, the answer might very well be "Yes." I was reluctant to lighten Kay's deep state of hypnosis as I felt I would be better able to probe her subconscious memories where she was. Kay is an articulate, well-spoken lady when not in hypnosis. Notice her abbreviated responses as we proceeded.

RTJ: And what country was that?

K: Egypt.

RTJ:	Where is the boat going, Ruth?
K:	Italy.
RTJ:	Why are you going to Italy in this boat?
K:	To get away.
RTJ:	And what are you trying to get away from?
K:	People.
RTJ:	Are these people trying to hurt you?
K:	They're trying to hurt many people.
RTJ:	Who are these people who are trying to hurt so many people?
K:	Soldiers.
RTJ:	Can you tell me what year is this, Ruth?
K:	Sixteen.
RTJ:	Is this 1600 or just sixteen?
K:	Sixteen.
RTJ:	Do you get to Italy?
K:	Yes.

After Ruth got off the boat, I took her forward in time a few years. Her father's name was Ben and her mother's name was Naomi, but she didn't know if they were alive or dead. She stated that she had no mate, no children, and no brothers or sisters.

I again took her forward until she was thirty years old, and in probing her lifestyle in Italy in the early Christian era, the following occurred:

RTJ:	Now I want you to go forward in time again. Go forward until you're about thirty-years old. Where are you now?
K:	Little village.
RTJ:	What country is this in?
K:	Rome.

RTJ:	Look down at your self. What are you wearing, Ruth?
K:	A loose dress, with a belt.
RTJ:	What color's the dress?
K:	Blue.
RTJ:	What's the dress made out of?
K:	Linen.
RTJ:	Look down at your feet. Are you wearing anything on your feet?
K:	Sandals.
RTJ:	And what kind of work do you do?
K:	Church.
RTJ:	What kind of a church is this?
K:	A Christian church.
RTJ:	Who is the ruler of Rome right now?
K:	Don't know. Too far away.
RTJ:	What's the name of the village where you are? Does it have a name?
K:	No. Not a town. Just a place.
RTJ:	Look at the houses in the village. What are they made of?
K:	Clay.
RTJ:	Now Ruth, I want you to see yourself. You're sitting down to a meal. Tell me, what are you eating?
K:	Bread. Cheese. Wine.
RTJ:	Where are you that you're eating this meal?
K:	Home.

Another literal answer. Again notice her one-word answers.

RTJ:	Is there anybody there having the meal with you?

K: No.

RTJ: Are you at a table? Are you inside or outside?

K: Table. Wooden table.

RTJ: Look around the room. What else do you see besides the table?

K: Fireplace.

RTJ: Are there eating utensils on the table?

K: No. Just plates.

RTJ: What are the plates made of?

K: Wood.

RTJ: Now Ruth, I want you to see yourself. You're at a market someplace. Tell me, what are you buying at this market?

K: Fish.

RTJ: And how do you pay for this fish?

K: Coins.

RTJ: Now I want you to see the coins. You're holding the coins in your left hand.

                Look very carefully. Describe the coins to me.

K: Uneven. A face. Some writing but I can't read it because I can't read.

RTJ: Look at the face real close. Do you know who that face is?

K: No.

I was having such good responses from Kay, I thought I would see if she would speak in another language.

RTJ: Now Ruth, there's somebody standing right next to you. They're a friendly person, and they're talking to you. What are they telling you?

K:	They're not talking to me. They're talking to someone else.

Notice here, she didn't respond directly to my suggestion that someone was talking to her, but corrected me to say that they weren't talking directly to her.

RTJ:	All right. What are they saying?
K:	Fish is too high.
RTJ:	Now listen very carefully to what they're saying. You're going to be able to repeat what they're saying in the same language. Now tell me, what are they saying, in the same language?
K:	(Kay struggles here. Her lips are moving.) Ik...ik...
RTJ:	I'm going to count from one up to three. When I get to three you'll be able to repeat exactly what they're saying. It will be very clear. One...two...three. Now it's very clear.
K:	Ik.... Can't hear. I can see but I can't hear.
RTJ:	Tell me what language is this that they're speaking?
K:	Latin.
RTJ:	Now I want you to go forward in time to some kind of a ceremony or ritual. Now tell me Ruth, what's the purpose of this ceremony?
K:	Baptism.
RTJ:	Who's being baptized?
K:	Friend. Lives nearby.
RTJ:	And what kind of religion is this?
K:	Christian.
RTJ:	What's your friend's name?
K:	James.

RTJ:	Now without any pain or discomfort, I want you to go forward to the day that you died in that lifetime. How old are you now?
K:	Thirty-seven.
RTJ:	Look around and tell me. Where are you?
K:	Cave.
RTJ:	Why are you dying in this lifetime?
K:	Just old.
RTJ:	Is there anybody there with you in this cave?
K:	No.
RTJ:	Why are you in a cave?
K:	No home.

The three past-life experiences that I have presented thus far were fairly ordinary, taking into consideration the apparent time-periods and locations in which the subjects lived. In their past-life personalities, Carol, Jane, and Kay were not prominent people in their communities even though their experiences were very interesting.

It seems to be a common belief that people who regress to a past life turn out to be very important people. However, in all of the past lives of my eighty-one subjects who regressed in this first research, I encountered no Cleopatras, Napoleons, high priestesses, or others who were prominent in their respective communities. If we *have* lived before this lifetime, there is probably someone somewhere who was President Lincoln, Cleopatra, or Napoleon in one of their past lives. But if you consider the very large number of us plain folks in the world, present and past, that I encountered only plain people in my subjects' past lives isn't surprising.

A valid observation, in view of these rather ordinary past lives, is that these experiences seem not to be mere wish-fulfillments of deeply-held personal desires, or projections of individual personal beliefs and religious backgrounds. Who, living in the United States in this time-period would wish to be a woman-of-the-streets in 1864 (Carol), or a pygmy living in the jungle (Jane), or living in a cave near Rome (Kay)?

Regardless of the ordinariness of their past lives, Carol, Jane, and Kay, as did many of my subjects, while experiencing what seemed to be past lives, appeared immersed in their experiences. They responded to my questions with facial expressions, emotions, and tones of voices that indicated they were reliving the experiences just as the memories were incorporated at the time the experiences actually happened, rather than merely reciting a story.

One of the most interesting parts of the past-life experience for many of my subjects, was what happened to them in-between lives. A frequent request was: "If we do this again, spend more time exploring what happened in-between lives." It was certainly new territory for me. I had no conception of what to expect. Would we encounter a blank, a vacuum, pearly gates, spirits floating on clouds, fire and brimstone, divine personalities, or what? Let's look at those experiences in the next two chapters.

4

Just Floating

Of the 104 subjects in my first research project, eighty-one regressed beyond birth to what seemed to be past lives. Of that eighty-one, approximately forty per cent died in those past lives of what apparently were natural causes, approximately nine per cent died violently, and about ten percent died accidentally. One died by committing suicide, and about five per cent died of what appeared to be disease. The causes of death of the rest are unknown.

One who died violently was Roger, a thirty-five-year-old white male. On his Information Survey Roger indicated "other" for his religious preference, and "slight" as his degree of involvement in religious activities. He had a Medium response to the susceptibility test and had a self-report depth of seven while in hypnosis.

Roger regressed in his first past life to being a nineteen-year-old man named Jonathan Morse, living on a farm near Atlanta, Georgia in the early 1860s. (In this life Roger was actually born and raised in Michigan.) He lived in a large stone house, with pillars, with a big, white front door. While in the house, he could see a large, curved stairway going up to the second floor. His mother was named Cynthia and he called his father Colonel David. Lincoln was President, "…but we aren't happy with him."

As we approached the day Jonathan died in that lifetime, the following occurred:

RTJ:	Just go forward then to the day you died. Can you see yourself on the day you died?
R:	Yeah.
RTJ:	All right. Where are you on the day you died in that lifetime, Jonathan?
R:	I'm under a tree.

RTJ:	You're under a tree?
R:	Scared.
RTJ:	About how old are you the day you died?
R:	I'm nineteen.
RTJ:	Can you tell me why you died? What was the cause of your death?
R:	I wouldn't fight.
RTJ:	Then why are you dying?
R:	The Captain is mad. I won't fight.
RTJ:	Now I want you to see yourself. Just after you died. Tell me, why do you think you died in that lifetime?
R:	The Captain shot me 'cause I wouldn't fight.

Usually the subjects would respond concerning their deaths in a manner that was consistent with the time-period and life style of the past-life personality, but which gave little concrete information. For example, a forty-year-old woman regressed in her first past life to being a brown-skinned woman named Marisa, in the 1800s. Why did she die? She responded: "Just old age."

A Thirty-year-old man regressed in his first past lifetime to being a French man named Michael, in 1750. The reason he died was "bad health."

A Thirty-two-year-old man, regressed in his first past life to being dressed in armor, carrying a spear and a shield. He died at age forty-five by being stabbed in the stomach by a large spear.

Whenever the subjects responded with an imprecise answer as to why they were dying, I listed it as "natural" in my statistics for lack of a better classification.

A past-life experience that I found especially interesting is that of a thirty-two-year-old woman of an oriental heritage. Her Survey showed that she had two years of college, was born and raised in California, and was married with two children. She regressed to being a thirty-five-year-old Scandinavian woman named Anisa, living in 1862, married to a man named Robert. When she spoke of Robert, she indicated great affection for him. "Oh…(lovingly) there's Robert."

RTJ:	Now I want you to go forward in time to the day that you died in that lifetime. You will experience no pain, no fear, and no discomfort.

Anisa sighed and by facial expressions and bodily movements indicates great discomfort.

RTJ:	If the experience is uncomfortable, we can go back up on the cloud…and you won't experience it.

After a short time, Anisa relaxes and becomes calm again.

RTJ:	Now I want you to go forward in time to the day that you died in that life time. Now on the day of your death…how old are you?
A:	I'm sixty.
RTJ:	You're sixty. And where are you Anisa? Where are you?
A:	I'm on a big boulder.
RTJ:	You're on a big boulder?
A:	Uh-huh.
RTJ:	I see. What was the cause of your dying?
A:	I jumped.
RTJ:	You jumped. Did you purposely jump? Or did you fall?
A:	No. I jumped.
RTJ:	All right. Now death is very near…very near. What have you been taught that happens after death?
A:	My life will end.
RTJ:	Now you will feel no pain…no discomfort…and you'll be able to tell me without any fear or discomfort. After you died in that lifetime. Can you see yourself after you died?
A:	I float up…I float up…float up.
RTJ:	You float up?
A:	Yes.
RTJ:	All right. Can you see other people where you died?
A:	They're concerned.

RTJ:	Is your husband Robert still alive?
A:	No.
RTJ:	He's not alive any more. Have you had any children by then?
A:	No.
RTJ:	Now, after you've died…after you've floated up. Where did you go…where are you going?
A:	I'm just…I'm just floating.
RTJ:	You're just floating?
A:	I don't know where I'm going.
RTJ:	Are there any other persons there…where ever you're going?
A:	No…I think…(suddenly expressing great surprise) Yes, there are…but I don't know who they are.
RTJ:	Can you see them…or do you feel them?
A:	I can see them.
RTJ:	Tell me what they look like.
A:	I can't quite make them out.
RTJ:	Can't quite make them out. All right.
A:	But I see colors…I can see colors.
RTJ:	Are you happy there? Do you feel comfortable there?
A:	Yes…but I'm afraid…happy…but afraid.

Anisa became very agitated at this point, so I took her back up on the cloud. Later the following occurs:

RTJ:	Now I want you to go forward in time. Just before you were born into another life time. Let me know when you're there by raising your right index finger that I'm touching gently now.

Subject raises her right index finger.

RTJ:	Oh that's fine. Now, are you choosing to be born?
A:	Yes.
RTJ:	What purpose...is there a purpose in your being born again?
A:	To help people.

Then there is Joe, a white man, fifty-two-years old. In his first past life he regressed to being a farmer named Carl, living in Wales, England, in 1742. He was married with two children. He died at age forty-eight.

RTJ:	And why are you dying?
C:	I worked too hard.
Later.	
RTJ:	Now without any pain or fear, I want you to pass right through that death experience. Now you've died in that life-time...tell me.... Now that you've just died in that life-time, can you see your body?
C:	Uh-huh.
RTJ:	Where are you that you can see the body?
C:	I'm still in the living room?
RTJ:	And where are you in the living room? Where are you that you can see the body?
C:	I've left, but I haven't.
RTJ:	Wherever you are, seeing the body. Is there anybody there with you?
C:	No.
RTJ:	Now how are you feeling? You just died in that lifetime.
C:	Like I've had a load lifted off of me.
RTJ:	Now I want you to go forward in time. Now you still haven't been born into another lifetime...but you've left that area where you died. Now tell me, where are you now Carl?

C:	I can't tell.
RTJ:	Is there anybody there with you?
C:	No.
RTJ:	All right.
C:	There seems to be a light.
RTJ:	There seems to be a light. Now look at yourself. Do you have a body of any kind?
C:	No.
RTJ:	Have you received any instruction during this period of time after you died?
C:	No.
RTJ:	And have you received any punishment?
C:	No.
RTJ:	Now I want you to go forward. You're just about to be born again. Now tell me, are you choosing to be born again?
C:	Yes.
RTJ:	And are you choosing what your sex will be in your next lifetime?
C:	Uh-huh.
RTJ:	And are you choosing your new parents?
C:	I think so, yes.
RTJ:	And is anybody helping you choose?
C:	No.
RTJ:	Tell me, why are you being born again?
C:	Seems like it's time to come back.

Another past life that I found of special interest was that of a forty-one-year-old lady named Mary. Mary is white, a college-graduate, born and raised in New York. She was brought up in the Jewish religion. She indicated no involvement in religious activities, but she reported that she believed in a life after death, and that she believed in rebirth after death. In the second lifetime to which she regressed, she was of a different sex and a different race, than in her present life.

RTJ: Now again, I want you to go forward in time...without any pain...without any fear or discomfort...go forward to the day that you died in that lifetime...and tell me, how old are you now?

M: I'm not old.

RTJ: Now why are you dying? What's the cause of your death in that lifetime?

M: A sword.

RTJ: Somebody kills you? Who was it that killed you, do you know?

M: No.

RTJ: Why did they kill you, do you know?

M: No.

RTJ: All right, now before you die, I want you to look down at yourself...and tell me...what clothes are you wearing? This is the day that you died.

M: They're black. There's nothing on the arms. Just black shorts.

RTJ: Are you wearing any kind of shirt?

M: Like a vest shirt.

RTJ: And look at your arms. What color is your skin?

M: Not deep-black, but it's not white.

RTJ: All right...and are you wearing anything on your feet? Look down, you can tell me.

M: They're chains...they're chains.

RTJ: Do you know why you're in chains?

M: No.

RTJ:	Now I want you to look back for a moment at all the people you knew in that lifetime. Do you know any of those people in other lifetimes?
M:	(Shakes head no.)
RTJ:	Do you know any of those people in your present lifetime?
M:	(Nods head yes.)
RTJ:	And who are they in your present lifetime?
M:	My father.

Subject is crying. Back up on the cloud.

The foregoing cases give you a glimpse into my subjects' encounters with both natural and violent deaths, feelings of relief after death, uninterrupted consciousnesses after death, knowing people from their past lives in their present lives, and the rebirth experience. In every one of the past lives experienced by my eighty-one subjects who regressed while in hypnosis, when they died, regardless of the cause, their consciousnesses continued on after their deaths. Of great importance, this surviving entity, whatever its nature, is not destroyed by our biological death.

I had carefully outlined the areas I wanted to explore before I started holding the hypnotic sessions. I expanded different areas of inquiry as I progressed, but had to cease certain other areas of inquiry, simply because I obtained no meaningful results. For example, I tried to inquire how long the subjects spent in-between lives, but the subjects consistently seemed to have no conception of time in that state. So much for some of my clever ideas.

In spite of not achieving everything I had planned on, look in Chapter Five at two most extraordinary in-between lives that I encountered.

5

He's Like Us

Immediately after both Anisa and Joe died (Chapter Four), their consciousnesses continued and, in some form, they just floated up out of their bodies. After *all* of my subjects died in their past lives, their consciousnesses continued on after their deaths.

It seems apparent that although consciousness itself is intangible, it is far greater than just a function of the physical brain, like awareness.

After *almost* all of my subjects died in their past lives they immediately in some form, just floated up out of their bodies. In those cases where the subjects did not report "floating up," they didn't report anything different. There were just complications involved. Some weren't visualizing details vividly enough to report anything. Some went right into another lifetime, and so on.

Some common examples of what my subjects recited as happening to them immediately after their deaths in past-lives are:

> "I jumped into the air," "I see myself rising out of bed," "I'm above my body—just went upwards," "just floating," "in the air," "in the sky" "floating near the ceiling," "I'm above my body," "just in space...it's soft there," and "going up—just a form."

Both Anisa and Joe chose to be born again and each gave a reason for being born again. Most of my subjects who regressed, who were asked, did choose to be born again. About twenty-nine per cent did not so choose but were born again anyhow. Interestingly, about seventeen per cent indicated they had help in making the choice to be reborn.

Some common examples of what my subjects recited as the reasons they are being born again are:

> "To learn," "to get it right," "I have to do something," "going back with friends—they're going back," "I have so many things to learn," "to finish up

unfinished business," "to have a better...different kind of life," and "I have to come back."

You will recall that Mary (Chapter Four) knew one person in her present lifetime that she had encountered in her past life. As with Mary, approximately twenty-six per cent of my subjects who regressed seemed to know people in their present lives that they had encountered in their past lives. Keep in mind that I attempted to take all my subjects back to two past lives of their own choosing. I didn't direct them into any particular life time or time period. It may very well be that other life times occurred in which they encountered the same people, over and over again.

I attempted to ask each subject immediately after they died in their past lives, if others were there with them, floating up above their bodies. About twenty-six per cent of my subjects did encounter others immediately after their death, and about twenty-five per cent did not. However, after I took them forward in time, still in their in-between lifetimes, some who didn't immediately encounter others did see or feel others later. Some examples are:

> "Don't see people—but feel them around me," "I sense someone or something there with me," "I can feel my Sarah," (the subject's wife who predeceased him in that lifetime), "others there—two or three—they feel familiar," "I feel a presence—don't know who," "others there—not people," "can't see them—but I know others are there-don't know them," "someone there-can't see—feels like it," "just sense others—can't see," "others there," and "friends there—they love me."

You will recall, that in both Anisa's and Joe's lifetimes, I asked them how they felt either immediately after death or sometime during the period in-between lifetimes. They both indicated some sort of relief. This type of response was almost universal with those whom I asked. Common responses were:

> "Feel at peace," "feel happy and free and loved," "feel happy and kind of sad," "just feel assured that everything is okay," "feel ready-don't know what ready for," "relieved—glad to be out of there," "feeling free," "feeling calm." However, one said she was "feeling scared—alone."

One other point of interest. With Joe, I asked him while in the period between lifetimes, if he had received any instruction or punishment. He replied that he had not.

All of my subjects who regressed, who were asked if they had received any punishment, responded that they had not. Several responded "no" to the question in almost a contemptuous tone of voice.

However, ten of my subjects indicated they had had some form of instruction or advice in their in-between life periods. Their responses were:

> "Seems like someone is giving instructions," "have counsels" (or possibly "have councils"), "Yes. I'm told I will live a life free of death—free of killing," "There is much to be undone," "I'm going to be born again—still have more to learn," and "Yes, by the wise people."

Two subjects, Sally and Carol, both had extraordinary in-between life experiences. Sally was a forty-six-year-old, white, single woman with two children. She disclosed on her Survey that her religious affiliation was Catholic, although she indicated no involvement in religious activities. She also indicated that she believed in some form of life after death, and that she believed in rebirth after death. Sally was an excellent subject and regressed to three lifetimes. First, she was an Indian woman named "Sun", wearing skin clothes and moccasins, in the 1200s. In her second lifetime, she was a ten-year-old boy named Tom, in 1856, working on a whaling boat.

Later at age thirty in that same lifetime, Tom lived in a small house in Winston (no country was ever specified). He had lost one leg by having it caught in a rope while whaling.

In her third lifetime, Sally was a white woman named Annabelle Atkins, living in a town named "Fairview" (no country determined), speaking the English language. At her death in that lifetime, the following transpired:

RTJ:	Now, I want you to go right through that death experience, without any fear now. Now you've just died. Tell me, can you see your body? Where are you?
A:	Standing with these tall people...very tall.
RTJ:	Do you know who they are?
A:	Robes on. I'm not as tall as they are. (Then expressing surprise) They're seven-feet tall.
RTJ:	Seven feet? How are you feeling?
A:	I feel okay.

RTJ:	Okay?
A:	They're guiding me.
RTJ:	Where are they guiding you to?
A:	They're telling me…I have to stay there for a while.
RTJ:	Why do you have to stay there?
A:	It's not time for me to go.
RTJ:	Do you know anybody that's there?
A:	There's…(with amazement) he's seven-feet tall and he has red hair…and he's telling me to stay.
RTJ:	And who is he? Do you know?
A:	He's kind.
RTJ:	He's kind?
A:	He's telling me to be patient.
RTJ:	Are you receiving any kind of instruction?
A:	They said I could work there…with them. I'm reading.
RTJ:	Are you receiving any kind of punishment?
A:	No.

I need to point out here, that it wasn't entirely clear to me if Sally was in-between lifetimes, or had gone immediately to another lifetime. That sometimes happened.

A little later.

RTJ:	Now I want you to go right back up on your cloud…where it's safe and serene…safe and serene. Now this cloud is going forward. It's starting to speed up. It's going forward…and it's bringing you back to the present time. But it's nighttime on the cloud. You can look up and see all the stars and all the planets as the cloud is bringing you back to the present time. Tell me now, have you ever lived on any of those other planets?

A:	(With surprise) That's where those people are. The tall ones.
RTJ:	Have you ever lived on another planet?
A:	I don't think so.

If you have followed the recent books and articles on the "near-death" experience, you will notice that the death experiences of my subjects in their past lives, and their experiences after death are quite different in most respects from those reported as the experiences of those near-death.

The near-death experience generally refers to the experience of a person who was apparently clinically dead, but then revived.

The common elements of these near-death experiences as they are reported, are that after "dying", the persons hear a loud ringing or buzzing noise; they experience moving rapidly through a long, dark, tunnel; finding themselves outside their physical body but being able to see their former body; then other disembodied people come to greet and help them. They then encounter a "being of light", some sort of a divine spirit, who shows them a review or playback of their lives.

As pointed out by Dr. Raymond A. Moody, Jr. in his 1998 book *Life After Life,* the above recitation is not necessarily any one-person's experience, but is a composite of experiences.[12]

It may not be an exaggeration to say that Moody's 1975 *Life After Life* became the definitive word on the actuality of the Near-Death-Experience to many who read his book. However, in his 1999 book *The Last Laugh*[13], Dr. Moody recites how the publishers of his 1975 book edited out his true opinions on the reported Near Death Experiences, and embarrassed him with "…untruthful exclamations like 'Scientific Proof of life after death'."

Past life researcher Dr. Ian Stevenson and two others were able to secure the medical records of a number of patients reporting near death experiences. They discovered that forty-five per cent of them did have life-threatening illnesses or injuries, but fifty-five er cent of them were rated as having had no life-threatening conditions.

They concluded "…that an important precipitator of the so-called near death experiences is the belief that one is dying—whether or not one is in fact close to death."

Neither Moody or Stevenson questioned the truthfulness of those who told the near death stories, they just questioned if they actually transcended their deaths into another realm, and then returned.

The near-death studies, like most studies, have strengths and weaknesses. The fact that a large number of persons who "died" and then came back, and had similar experiences, certainly warrants investigation. The weaknesses however are several and serious. These persons didn't really die in the usual sense, in that they didn't stay dead. The studies of the experiences are really just studies of anecdotal stories, expressed by persons in terms of their personal beliefs and religious backgrounds. (How else could they be expressed?)

For example, many of the experiences reported by the persons during near-death, include seeing earth-like scenes after they left their bodies. Some saw gardens, green grass, flowers, and beautiful buildings. Some saw religious figures such as Saint Peter, God, Christ, Hindu messengers of death called Yamdoots, the Hindu king of death Yamaraj, and some heard Angels singing. These would seem to be cultural artifacts. In addition, they encountered long-dead relatives and others, who were never reborn again into new bodies.

Researchers Melvin Morse and Paul Perry in their 1990 book *Closer to the Light*,[15] believe they have discovered (or possibly re-discovered) the neurological explanation for the near-death experience. They point to an earlier work of Dr. Wilder Penfield that reported that electrical stimulation near the area of the brain known as the Sylvian fissure, in the right temporal lobe located just above the right ear, caused out-of-the-body experiences in which the patients saw God, heard beautiful music, and saw dead friends and relatives.

In preparing for my research, I determined that a study that involved the recitation of the actual experiences as they were apparently encountered would give more accurate information as to what actually happened; rather than a collection of anecdotal stories expressed in terms of the subjects' current world-views.

None of my subjects had contact with a divine spirit who conducted a review of the events of their lives. *Almost*, I emphasize *almost*, none of my subjects had contact with any type of divine being. One of my subjects did have contact with a form with a female face that she described as an Angel. Then, of course, there were the seven-feet tall beings encountered by Sally, whatever they might have been.

However, in addition to Sally's experience, the second very unusual in-between lifetime experience was that of Carol's. Carol was a married lady with two children, in her mid-forties, well educated with graduate degrees, raised in middle America in a Protestant home. She indicated that she was moderately involved in religious activities, believed in some form of life after death, and was uncertain as to her belief in past lives or rebirth. She was one of the few that I regressed to more than two lifetimes.

In Carol's first reported past life (Chapter Two), she lived as Maria in 1864, a lady of-the-streets, probably in Italy. As we approached her death in that lifetime, let's continue on:

RTJ: Now you've just died in that lifetime. Can you see your body?

C: Yeah.

RTJ: Where are you that you can see your body?

C: I'm looking down.

RTJ: Is there anybody else up there with you?

C: I can see them all there.

RTJ: Who are they that are up there with you?

C: All the people from the town.

RTJ: Do you know any of those people that are up there with you?

C: They're family that went on before. Other people that I knew.

RTJ: Why are they there? Did they come for any purpose?

C: They came...to welcome me.

Notice this is the only instance in my cases where other discarnate personalities come to welcome the deceased, similar to the near-death experiences.

RTJ: Now go forward just a little bit. What are you doing now? Are you still there looking at the body?

C: No.

RTJ: Where are you now?

C: Sometime I'm with others when they go, and some are coming in.
            Sometimes I'm just resting...being tired.

RTJ: Look down at yourself. Do you have a body wherever you are?

C: Hmmm. I never thought of that before. Yeah (with surprise).

RTJ: What's it look like?

C: It's kind of a…kind of like a human…the texture's not the same…but you know who you are…and who the others are.

RTJ: Are you wearing any kind of clothes?

C: No.

RTJ: Does your body look just about how it did when you died?

C: No…it's not the same. It doesn't have hair and all that kind of stuff. It's more like…who you are.

RTJ: Are you receiving any kind of instruction? Do you have a greater understanding of things there?

C: Yes.

RTJ: Where did you get this information? Where did this greater understanding come from?

C: Sometimes it just comes on…sometimes…ah…sometimes just when you leave your body…sometimes there's God…and he lets you touch him.

RTJ: Can you see God?

C: Yes.

RTJ: What's he look like?

C: He's like…light and warm…and…he's like us. You know there's no skin or stuff like that anymore. (With emphasis.) He's just there…and you're just joining…and when you touch…you get some more understanding…and he takes away some of the hurt…the pain from wherever you're been. (With finality.) He just touches and it happens.

The experiences of my subjects in many respects were remarkably similar to the experiences of subjects recounted by Dr. Edith Fiore in her excellent book *You have Been Here Before*.[16] Dr. Fiore's subjects also experienced that their consciousnesses continued immediately on after their deaths in past lives. Her subjects experienced a release from the pains and troubles of their past lives, as did my subjects. Her subjects almost all used the term "floating" immediately after their deaths. My subjects used either the same or similar expressions.

The similarity of these many cases where the consciousnesses of the deceased individuals just "floated" up out of their bodies, both with my subjects and those of Dr. Fiore's, almost none of whom knew each other or had the opportunity to become aware of each others' experiences, certainly lends great credibility to the authenticity of the event. To suggest that the similarity of these experiences in just "floating-up" after death comes from some form of communication between all the subjects, would have to assume some kind of conspiracy between strangers of a very great magnitude.

My purpose in reviewing the near-death studies is not in any sense intended to fault the methodology of those researchers. They had to work with what was available to them to study. Those were the stories of those who "came back", told in terms of their personal beliefs and religious backgrounds.

In the next chapter, I want to explain the process of using hypnosis in past-life regressions, review some of the common misconceptions concerning hypnosis, and most importantly, try to explain what occurs in the trance state so that the reader can put these past-life experiences into perspective, evaluate their validity, and use them in evaluating the hard core cases published by other researchers.

I will also explore the past life of a woman who regressed to being a man who was hung for murder in 1792.

6

I'm a Killer

As a further aid to evaluating what seems to be my subjects' past lives and the experiences of others, I would like to briefly discuss the use of hypnosis in conducting this type of research.

Many people, possibly most, tend to think of hypnosis in terms of the occult, the bizarre, or some form of mind control. Actually, the hypnotic trance state is a voluntary one. I can't force a subject into a trance. I can only lead a subject there if the subject allows me. Hypnosis does not involve overcoming the person's "will". If the subject had any reason not to go into a trance and "willed" not to, he or she simply just wouldn't go.

Hypnosis involves no mind control. If I suggested to a subject in hypnosis that he should go rob a bank and bring me the money, or if I suggested anything that would violate her moral or ethical standards, the subject would either come right out of the trance, or just ignore such inappropriate suggestions.

Also contrary to common belief, a subject is not asleep or unconscious while in a trance. If I touched a subject while not in a hypnotic trance, the subject would of course be aware of it. If I touched a subject while in a trance, the subject would also be aware of it.

Hypnosis is increasingly being used by sophisticated psychotherapists to quickly reach back into the subject's past to uncover trauma, defenses, and behavior patterns; but hypnosis alone is not therapy.

The discipline is centuries old, not a part of the "New Age" as some claim. However, the period since World War II has been the golden age of research into this phenomena. During this period, hypnosis has become a science, with a large body of knowledge being accumulated by serious and qualified researchers. It is also an art. There is a subjective-skill involved in inducing, deepening, and working with trance-states. Skills vary from person-to-person, just as in other disciplines.

Some excellent schools are available at which you can learn these skills. In California, for example, the training courses of the better schools and the instructors are approved by the Superintendent of Public Instruction. Unfortunately, some very inadequate schools are also around.

In spite of the extent and quality of the research, no specific definition of hypnosis can be agreed upon by everyone. However, I believe many hypnotists would generally agree, that hypnosis is an altered state of consciousness, characterized by focused attention, relaxation, and heightened awareness.

The different depths of hypnosis, frequently, but not always, result in different behavioral-responses. Numerous attempts empirically to measure these depths have been made, resulting in a number of numerical indexes and scales. Most of these indexes and scales are fairly useless in anticipating what an individual subject will do. Each person is unique and responds differently to hypnosis and to the hypnotist.

In addition, almost every subject will fluctuate in depth from time-to-time during each individual session. The competent hypnotist will generally be able to monitor and be aware of these changes as the session progresses.

My opinion, based on my experience, and the opinion of many hypnotists with whom I have discussed it, is that in age-regression (past-life regression is a form of age regression) the best results are generally obtained where the subject is in at least a medium trance-state, and preferably in a medium-deep state. Because of the unique nature of each subject, however, this isn't always true.

In attempting to understand what happens in past-life regressions while subjects are in hypnosis, "trance logic" deserves special attention. In hypnosis, a dissociation occurs; a partial separation between the conscious and subconscious mind. (Psychologists would generally refer to what I call the subconscious mind as the unconscious mind.) Once the trance state has been induced and deepened to a sufficient depth, the hypnotist is interacting primarily with the subconscious mind. The importance of depth in a trance, is that in a light trance, a subject can analyze, judge and/or distort information, depending on some agenda of their own, if such be their wish.

The hypnotized subject can, after such dissociation has occurred between the conscious and subconscious minds, accept and deal with, logically-inconsistent perceptions of reality. The subject's on-going reality-testing is reduced to the point where the subject is mostly oriented to internal experiences. The subconscious mind can then accept whatever reality is provided for it, as long as what is provided is not offensive to the subject.[17]

For example, in my research, after inducing a medium or medium-deep trance state, I had the subjects see a large white cloud. I had the subjects climb up on the cloud; and the cloud then takes them back through time and space to a time when they were much younger; and then back before birth.

By ordinary logic, of course, traveling on a cloud makes no sense. In trance logic, this anomaly is accepted without difficulty.

With this background in mind, let's look at the past-life regression of Sarah.

This regression is a very good example of what hypnotists call "revivification." The subject is immersed in her experience, seemingly reliving the experience exactly as the memory was incorporated. Unfortunately, the emotions, facial expressions, inflections, timing, pauses, and tone of voice doesn't come through in a verbal transcription.

Sarah is a single, white woman with three children, in her late fifties. She is a college graduate, employed in the scientific field. On her Information Survey she indicated "other" as her religious affiliation, and marked "uncertain" in her answers to my questions, "Do you believe in some form of life after the death of the body?", and "Do you believe in reincarnation?" Her self-report state in hypnosis was seven.

After easily placing her into a deep state of hypnosis, I placed her on a cloud upon which to travel, to go back into time and space. I then regressed her back to the age of fifteen, then five, and then to a time before being born into this lifetime.

She signaled me that she had arrived at a past life by raising her right index finger as I suggested she would do.

RTJ:	Now without any pain, fear, or discomfort, come down from your cloud.
	You're going to be able to talk to me in English or any other language that you may encounter. Now tell me, how old are you?
S:	I'm forty-seven.
RTJ:	And are you a woman or a man?
S:	I'm a man.
S:	And what's your name?
S:	My name is Ronald.

As Ronald, Sarah's responses were direct, matter-of-fact, and unemotional with little or no facial expressions. Ronald answered my questions, but volunteered nothing.

RTJ: I want you to look down at your arms. Now tell me Ronald, what's the color of the skin on your arms?

S: My skin in white.

RTJ: And look down at yourself, and tell me Ronald, what are you wearing?

S: Rags.

RTJ: Look down at your feet. Are you wearing anything on your feet?

S: No.

RTJ: What's your last name?

S: Aires. (Phonetic.)

RTJ: Look around Ronald. What do you see? Where are you?

S: Prison.

RTJ: Where is this prison located? What country is this prison in?

S: England.

RTJ: What year is this?

S: 1792.

RTJ: Look around you. Is there anybody close to you that you know?

S: No.

RTJ: What part of the prison are you in? Where are you in prison?

S: Death row.

RTJ: Why are you on death row?

S:	(Now, with a contemptuous look and tone of voice.) I'm a killer.
RTJ:	Who did you kill?
S:	I killed a man who tried to steal from me.
RTJ:	Did you know who this man was?
S:	No.
RTJ:	How did you kill this man?
S:	I hit him.
RTJ:	Did you hit him with anything, or was it just with your hands?
S:	I had a big club.
RTJ:	Tell me Ronald, do you have a mate?
S:	Not any more.
RTJ:	Did you use to have a mate?
S:	Yes.
RTJ:	What was her name?
S:	Her name was Elizabeth.
RTJ:	Did you and Elizabeth have any children?
S:	No.
RTJ:	What was your father's name?
S:	My father's name was Arnold.
RTJ:	What was your mother's name?
S:	Marina.
RTJ:	Now without any pain, fear or discomfort, I want you to go forward to the day you died. How old are you now?
S:	Forty-seven.
RTJ:	How did you die?

S:	By hanging.
RTJ:	Now without any pain, I want you to pass right through that death experience. Now you've just been hung. Tell me Ronald, can you see your body?
S:	Yes.
RTJ:	Where are you that you can see the body?
S:	I'm right up above it.
RTJ:	Is there anybody up there with you?
S:	I think so.
RTJ:	Can you see anybody up there with you?
S:	No.
RTJ:	How do you know there's anybody up there with you?
S:	I just feel it.
RTJ:	Now I want you to go forward in time. You still haven't been born into another lifetime. Now tell me, where are you now?
S:	It's very cool here.
RTJ:	Now look around, What do you see?
S:	I see light. I see color.

Several of my subjects saw colors in-between lives, whether or not they were aware of others there.

RTJ:	Do you see other people there?
S:	I sense people.
RTJ:	Do you sense anybody there that you know?
S:	I can't tell the difference between know and not know.
RTJ:	Tell me, have you received any instruction during this period between lifetimes?
S:	Yes.

RTJ:	What kind of instruction have you received?
S:	That I will live a life free of death. Free of killing.
RTJ:	Who gave you this instruction?
S:	I don't know.
RTJ:	Have you received any punishment during this time between lifetimes?
S:	No.
RTJ:	Now look back at the life you just lived. Do you know any of those people in other lifetimes?
S:	There does seem to be a knowing.
RTJ:	Do you know any of those people in your present lifetime?
S:	I'm not sure, 'cause it feels very comfortable.
RTJ:	Now I want you to go forward again. You're just about ready to be born into another lifetime. Now tell me, are you choosing to be born again?
S:	Yes.
RTJ:	Are you choosing what your sex will be in your next lifetime?
S:	Yes.
RTJ:	Are you choosing your new parents?
S:	No.
RTJ:	Is anybody helping you choose to be born again?
S:	Yes.
RTJ:	Who's helping you choose to be born again?
S:	A very powerful form of energy.
RTJ:	Tell me, why are you being born again?
S:	I have lessons to learn. I have things to do.

I then took Sarah back up on her cloud and took her back to another lifetime. In this second lifetime, Sarah regressed to being an Apache woman, and she seemingly spoke a language other than English.

Occasionally, in a past-life regression, the situation in which the subject finds herself/himself is so traumatic, that if not properly dealt with, it can be very upsetting.

One subject, upon being brought down from his cloud, looked around and said, "My God, they're all dead." He opened his eyes and looked around the room in terror. He had descended into a town in Poland during World War II. At my suggestion he closed his eyes, and went back up on his cloud, and the trauma was dealt with.

For this reason, I would caution those of you who might be using self-hypnosis for any reason, not to attempt to regress yourself. I would also suggest not allowing anyone to regress you who has not had training and experience in dealing with a traumatic situation that you might encounter while in hypnosis.

In the next chapter, let's look at Sarah's tragic second past life as an Apache Indian

7

Mommy, Don't Leave Us

I have seen no credible evidence to substantiate the criticism made by some, that the hypnotic subject, while regressed to what seems to be a past life, is merely fantasizing for romance to liven up an otherwise drab life. Such criticism seems to be just speculation. My experiences are not supportive of this particular criticism.

In my first research project, the post-hypnotic interviews, sixty-three indicated they had no knowledge of the past-life periods to which they regressed. Most of the rest indicated only a slight knowledge of the time periods. Only one subject had a past life where romance was involved. She was the mistress of a desert sheik but no erotic details materialized in her regression.

Millions of people world-wide,—probably more than those who subscribe to traditional Christian beliefs, believe in reincarnation. The doctrine isn't strange here at home. A Gallup poll a few years ago found that twenty-three per cent of the adults in the United States believed in the rebirth of the soul into another body, or in reincarnation.[18]

Suffering, fear, and tragedy seemed to besiege Sarah in her second past life. As we proceeded, Sarah seemed to be reliving her experiences, and she seemed to become more and more involved and preoccupied with what was going on. Many times, the subject is so preoccupied with the personality to which they have regressed, that my questions to them seem unimportant. When I asked her about her mate, for example, the important thing to her was the fact that he had gone off with the other men in the tribe, fighting their attackers.

RTJ: Now without any pain, fear or discomfort, come down from your cloud.

                You can talk to me in English or any other language that you might encounter. Now tell me, how old are you?

S: I'm a woman. (Obviously unresponsive.)

RTJ:	How old are you now?
S:	I have many children. (She apparently didn't understand the concept of age in years.)
RTJ:	Many children. All right, do you have a mate?
S:	Yes.
RTJ:	What's your mate's name?
S:	(Silence. Sarah seemed preoccupied here.)
RTJ:	Now you can see yourself. You're standing right next to him. What name do you call him?
S:	He's gone.
RTJ:	Now tell me, how old are you now?
S:	I don't know.
RTJ:	Look down at your arms. Tell me, what color is your skin?
S:	It's brown.
RTJ:	What's your name?
S:	(Silence. Sarah seemed very preoccupied here. I'm not even sure she was aware of my question.)
RTJ:	Now there's somebody standing right next to you. They're a friendly person and they're talking to you and calling you by your name. Now tell me, what do they call you?
S:	Prokeno. (Phonetic.)
RTJ:	What race are you Prokeno?
S:	Apache.
RTJ:	What year is this?
S:	I don't know year.
RTJ:	How many children do you have Prokeno?
S:	I have seven.

RTJ:	Tell me, what's the name of the oldest child?
S:	Racine.
RTJ:	Is this a girl or a boy?
S:	A boy.

A frequent criticism of using hypnosis to age-regress a subject is that the subject, being in a very suggestive state, is merely responding to the suggestions of the hypnotist. Notice here and in the following dialogue, Sarah frequently does not respond to my questions, but instead tells me what seems to be going on.

RTJ:	Now look around and tell me, can you see any dwellings…any kind of a dwelling that people live in?
S:	We're in the fields. We're traveling across the country.
RTJ:	Are there other people there with you?
S:	Yes. There's a great deal of fear and hatred.
RTJ:	Why are you afraid?
S:	Because other people are attacking us.
RTJ:	Look around and tell me now, do you see any animals?
S:	We have dogs.
RTJ:	Look down at your self. Tell me, how are you dressed?
S:	I'm wearing animal skins.
RTJ:	Are you wearing anything on your feet?
S:	Yes.
RTJ:	What kind of weapons are you carrying?
S:	I have a knife.
RTJ:	What's the knife made of?
S:	The knife is made of stone.
RTJ:	Do you have any other weapons with you?
S:	No.

RTJ: I want you to look around at some of the men traveling with you. Are they carrying any weapons?

S: No men. They went off to fight.

RTJ: Who are the men fighting?

S: The Others.

RTJ: Now I want you to go forward in time to the next significant event. Tell me, what's happening now?

S: I'm looking at my hungry children.

RTJ: Where are you?

S: There are mountains...and cold. We need food. We have no men...just women and children.

RTJ: Now I want you to go forward...without any pain or fear...forward to the day that you died in that lifetime. Now tell me, how old are you now?

S: I don't know.

RTJ: Look around and tell me. Where are you?

S: I'm very hungry.

RTJ: Look around and tell me, where are you?

S: In strange country...in mountains. It's very cold.

RTJ: Is there anyone else there with you?

S: No. I've gone by myself.

RTJ: Now without any pain or fear, I want you to pass right through that death experience. Now you've just died. Can you see your body?

S: Yes.

RTJ: Where are you that you can see that body?

S: I'm above it.

RTJ: Now I want you to go back a little bit before you died. You're with the women...you're still trying to get away

	from the battle. Now tell me,there's someone standing right next to you (subject sobs) and they're talking to you. What are they saying to you?
S:	They're children. (With anguish.) They're hungry.
RTJ:	Now tell me what they're saying in the same language that they're speaking. You can hear it now. It's very clear.

Here the subject seems to speak a few words in another language. I'm unable to even phonetically transcribe it.

RTJ:	What does that mean in English?
S:	Mommy, don't leave us. (With emotion.)
RTJ:	Now I want you to go forward a little bit…just before the time you went off by yourself. Are you telling anybody that you're going to go off by yourself?
S:	No.
RTJ:	Now without any pain or fear, I want you to go right through that death experience. You're right above the body again. Tell me, is there anybody there with you?
S:	Don't think so.

Sarah seemed quite agitated at this point, so I took her back up on her traveling cloud.

RTJ:	Now it's safe and secure on the cloud. Tell me, how are you feeling now that you've just died in that lifetime?
S:	I feel safer.
RTJ:	Look down at all those people you knew in that lifetime. Do you know any of those people in other lifetimes?
S:	No.
RTJ:	Do you know any of those people in your present lifetime?
S:	No.

Notice several things of interest here that we haven't seen in other past lives. When I ask her what year it was she replied, "I don't know year." Not, I don't know what year it is, just "I don't know year." I struggled throughout her regression to find out how old she was. As with most researchers, I became more sophisticated as I progressed. Obviously, I could have ask her how many winters she had seen, or similar approaches. I did do better later.

To some extent not knowing the time period to which the subject would regress, was a constantly occurring problem. If I had had some idea of the experienced time-periods, I could have done some research and come up with more definitive questions to try to establish the names of rulers and other pertinent information. This factor of not knowing what was coming next, however, added excitement and wonder to the research.

Many of the subjects, such as Sarah, seem to "relive" the experiences they were describing. This became apparent from time-to-time, not only by their facial expressions and emotions, but by their volunteering things that I hadn't asked for.

For example, Martha, a sixty-two-year-old woman, who first regressed to being a fifteen-year-old male living in Denmark in 1864. Her second past life was as a woman, living in England in the early 1800s. At age twenty-five she had a mate named Peter, and a young son.

RTJ: Do you and Peter have any children

M: I see a little boy.

RTJ: What's your little boy's name?

Here, instead of directly answering my question, she turned her head downward, smiled broadly and said:

M: He's pulling on my dress.

Again, there is a thirty-five-year-old man, who in his first past life regressed to being a fifteen-year-old boy named Frank. He lived on a farm. After establishing race, and other information, I asked the following:

RTJ: Now look down at yourself again, Frank. What are you wearing?

F: Overalls and a shirt. (And then he volunteered), My sleeves are rolled up.

RTJ: Are you wearing anything on your feet?

F: Work boots. Kind of a...yellow leather. (Then with
                      disgust.) Don't much like 'um.

A thirty-nine-year-old woman, who in her first past life regressed to being a
thirty-two-year-old woman, was living in Kansas in 1836. When I asked her to
look around and tell me where she was she replied, "In a cabin." And then she
said with a pained expression, wrapping her arms around herself, "It's so cold."

A woman subject regressed to being a man named John living in Nebraska in
1894 in a cabin. When I ask him to look around in the cabin and tell me what he
saw, he replied, "There's a stove, table, four chairs," and then he stopped and
acted as though my question had offended him. "It's my home."

Another woman, forty-four years old, regressed to being a woman in England
in 1858. Then in her second past life she regressed to being a woman in a wealthy
family, in France. After establishing her age, I ask her if she was a man or a
woman. She replied with intense feeling, "I'm a lady."

When the subjects "relive" their past-life experiences, their recitations and
their concomitant emotions make what I am hearing and seeing seem very real.
When a subject jumps off a cliff to commit suicide, or is stabbed, or is run
through with a sword, or dies with great trouble breathing, the emotions generated by the subject seem very authentic. Regardless of whatever *we* may think is
occurring in a past-life regression, in the subjects' minds the events described are
very real. It doesn't seem as if it were wishful thinking or role playing.

The suggestion has been made that what we may be dealing with here is Carl
Jung's collective-unconscious doctrine. Jung advocated that a part of our unconscious mind is inherited, is common to all persons, and is the seat of inherited
ideas and predispositions. However, Jung's primordial images would not seem to
be the source of the detailed past lives that my subjects exhibited—specific
names, places, families, and events.

Similarly, other types of genetic-inheritance theories would seem not to be an
explanation. As with Sarah, others also spoke of events that occurred after their
deaths, and after they could have propagated children to carry on the claimed
genetic memories. Deceased people just don't procreate.

You will notice that Sarah seemed to speak in another language. Actually, she
only spoke a few words. In the next chapter, we'll encounter a regression where
the subject seemed to speak another language at some length.

8

The Fishing is Not Good Today

You will recall that Jane (Chapter Three), in her first past life, regressed to being a pygmy named Inkwa, living in the forest. She died in a canoe accident in that lifetime.

After Inkwa had died, I inadvertently caused some confusion that resulted in an interesting situation. As I did with all my subjects who regressed, I tried to take Jane forward in time after she died as Inkwa, to explore her in-between life, but I took her too far forward.

Jane was in a deep state of hypnosis and seems totally absorbed in her past-life personality. Follow what occurred after she died but she was still in the personality of Inkwa:

RTJ:	Now tell me, where are you now that you've died?
J:	I'm in the tall trees.
RTJ:	Can you look around and see the forest wherever you are?
J:	I can see my village.
RTJ:	Is there anybody there with you?
J:	No. They're not with me. My body's lost. I'm in the trees.
RTJ:	Now I want you to go forward in time...forward in time to just about the time you're ready to be born again. Now tell me...where are you?
J:	I'm in my mother's belly.

Here I had taken Jane, as Inkwa, a little further forward in time than I had intended.

RTJ: Go back in time just a little bit…before you got in your mother's belly.

Tell me, where are you now?

J: It's a village.

RTJ: It's a village?

J: (With surprise.) There's big animals…I've never seen.

This obviously wasn't the village she lived in as Inkwa. She would have been familiar with the animals in that village.

RTJ: Is this another lifetime?

In going forward and back in time, occasionally the subjects go right into another lifetime.

J: It's not the same village.

RTJ: How old are you in this village with these big animals?

J: I'm not born yet.

At this point, I concluded that I had confused the situation, so I decided to go forward right into another lifetime.

RTJ: Now go forward again, you've just been born into another lifetime. Tell me, did you choose to be reborn again?

J: Yes.

RTJ: Did you choose your sex?

J: Yes.

RTJ: Did you choose your new mother and father?

J: Yes.

RTJ: What's your new mother's name?

J: Just…the lady.

RTJ: Now go forward in that lifetime. Go forward until you're just about ten years old in that lifetime. Now are you a boy or are you a girl?

J:	I'm a boy.
RTJ:	Tell me, what's your name?
J:	Zeehon. (Phonetic.)

Notice here, Jane's second past-life personality is that of a male and an oriental. She is Caucasian in her present lifetime.

RTJ:	And what's your mother's name?
J:	She's dead. Just…the lady's dead.
RTJ:	Is your father alive?
J:	I don't know who my father is.
RTJ:	Now Zeehon, I want you to look around. What do you see? Where are you?
J:	At the village.
RTJ:	What's in the village? Look around and tell me.
J:	Houses made of…(with surprise) bamboo?… And big fat animals. (then exclaiming)…Pigs.
RTJ:	Pigs.

When the personality Inkwa saw the big animals in the strange village, she didn't know what they were. Now, after being reborn as a boy named Zeehon, a new and different personality, she recognized the big animals as pigs. Jane was raised in rural middle-America and, of course, knows what a pig looks like. What I found to be most interesting here was that Jane's normal, present-life awareness, which would recognize a pig, has been completely pushed aside. I was interacting first with the personality Inkwa, and second with the personality Zeehon, neither of which were in contact with each other, or with Jane's normal awareness.

Now, continuing with the personality Zeehon:

J:	And we're on the river. The village is on the river.
RTJ:	Does the village have a name?
J:	Hmmm.

RTJ: I want you to see yourself talking to somebody...and you're talking about your village. Do you call the village by a name?

J: The village...at the river.

RTJ: Now I want you to see somebody standing right next to you...and they're talking to you. Tell me what they're saying? Repeat to me in the same language that they're speaking.

Here Jane speaks several sentences in what sounds like an oriental language.

RTJ: All right. That's fine.

J: The old man.

RTJ: It's the old man? What does that mean in English?

J: Just...the fishing is not good today.

RTJ: Now tell me, are there any other villages close to the village that you're in?

J: They're inland. It's not safe on the river.

RTJ: Now I want you to go forward in time...until you're about fifteen or sixteen-years old. Now look down at your arms. What color is your skin?

J: Brown.

RTJ: Now look at someone close to you. Look at their eyes. What do their eyes look like?

J: They're dark.

RTJ: Are they round or are they slanted?

J: They're slanted. Their hair is black. My hair is black...straight black.

RTJ: Now I want you to go forward in time...up until you're about twenty-five or thirty years old. Tell me, do you have a mate?

J: Yes.

RTJ:	And what's the mate's name?
J:	Oseeta. (Phonetic.)
RTJ:	Do you have any children?
J:	No.
RTJ:	Do you ever go to a market to buy anything?
J:	Uh huh.
RTJ:	Now I want you to see yourself going to a market. What do you use to purchase things with? Do you use money?
J:	We barter. We have coins on a string. Mostly we barter.
RTJ:	Look at one of the coins. Describe what this coin looks like.
J:	It's round...and dark...and it has a hole in the middle.
RTJ:	What shape is the hole in the middle of the coin?
J:	It's square.

A trip to the American Numismatic Association's museum here in Colorado Springs disclosed that some of the old Chinese coins do indeed fit Jane's description.

RTJ:	Now I want you to see yourself at some sort of a ceremony. Some sort of a ritual or ceremony. Now tell me, what kind of a ceremony is this?
J:	The yellow moon.
RTJ:	And what's the significance of this ceremony? Why are you having this yellow-moon ceremony?
J:	When winter begins.

A local retired Chinese physician, born, raised and educated in China, informed me that in some rural areas of China, there is a traditional celebration at harvest time called the Yellow Moon Celebration. The women bake small "moon cakes" for the occasion.

RTJ:	Do you have any children?

J:	No.
RTJ:	Do you have any leaders or rulers in your village?
J:	The old men.
RTJ:	Do you have any leaders or rulers in the country that you live in?
J:	They come down the river. They make us pay. But most of the time we don't see them.
RTJ:	Do they have a name? The ones that come down the river and make you pay?
J:	Chink See. (Phonetic.)
RTJ:	Now go forward to the day that you died. Tell me, how old are you now?
J:	Very old.
RTJ:	Can you tell me in years how old you are?
J:	65.
RTJ:	Look around you. Is there anybody there with you?
J:	There's the servant girl.
RTJ:	What's her name?
J:	Keena. (Phonetic.) I'm in bed...I have a long white beard.
RTJ:	Why are you dying? Do you know what caused your death?
J:	I'm sad.
RTJ:	And why are you sad, Zeehon?
J:	Oseeta is dead. I'm all alone. I'm old and I can't do things.
RTJ:	Now pass right through that death experience. You've just died. Can you see yourself?
J:	I'm in bed.

RTJ:	And where are you that you can still see yourself?
J:	Up above the bed.

Here we encounter what appears to be an unusual custom of body disposal after death.

RTJ:	Now I want you to go forward in time a little bit. Tell me, where are you now Zeehon?
J:	I'm on a boat.
RTJ:	Are there other people there with you?
J:	No.
RTJ:	You're all by yourself on this boat?
J:	Yes.
RTJ:	Where is the boat?
J:	It's a reed boat.
RTJ:	Is this your body?
J:	Yes.
RTJ:	Are you still in your body?
J:	No. I'm dead.
RTJ:	And where are you that you can see your body?
J:	Above the river.
RTJ:	Is there anybody there with you?
J:	Not on the boat. On the shore.
RTJ:	Is there anybody up above where you are...looking down?
J:	No.
RTJ:	Now I want you to go forward in time. You still haven't been born again.
	Look around and tell me, where are you?
J:	Above the clouds.

RTJ:	Is there anyone there with you?
J:	Oseeta.
RTJ:	Look at Oseeta. Does she have a body? Can you see what she looks like?
J:	No. I just know it's her.
RTJ:	Now have you received any instruction during this time after you left your body?
J:	No...We just have counsel.
RTJ:	Have you received any punishment?
J:	No.

I have made preliminary attempts to determine if the words Jane spoke as Zeehon are actually intelligent words in another language, possibly Chinese. To my dismay, I find that there are a very large number of closely-related languages exist called dialects, that are spoken in China, along with various aboriginal languages.

I first asked my retired Chinese physician acquaintance to listen to Jane's words. He said it wasn't a dialect that he knew, but thought it might be Cantonese. I later met a gentleman from China who spoke Cantonese and he said it wasn't that dialect. In both cases, I asked these two gentlemen if the words sounded like a language, or just gibberish. They both said it sounded like a language.I will pursue this, but it may be a lost cause.

In Chapter nine, I want you to meet Carol again. Carol is a most remarkable subject as we will discover.

9

We're all Going to Die

We first met Carol as a lady-of-the-streets in Chapter two, and again in Chapter Five where we explored her in-between lifetime, where after her death, she touched God.

As I mentioned in Chapter two, Carol is a married lady with children, in her late-forties, well-educated with graduate degrees.

Carol was, and is, a somnambulist; meaning that she is capable of readily going into a very deep state of hypnosis. So deep, that in her first session, I had to bring her up to a lighter state so she could speak more clearly.

Somnambulist subjects are exciting, not only because they are capable of many different responses, but also because the hypnotists feel that they are probably probing the depth of the subjects' subconscious minds.

In the post-hypnotic interview after the first session, Carol disclosed that one of the reasons she wanted to participate in my research, was because of some unpleasant and disturbing feelings and thoughts she had from time-to-time. These feelings were usually brought on by hearing about concentration camps, people dying and being hurt, or being subjected to out-of-control rage. Although hearing of such things would be unpleasant to most of us, she felt her emotional reaction to them seemed much more disturbing than it normally should be.

Because I wanted to work with her again (you don't encounter a somnambulist every day), I suggested a session where we could explore the origin of these feelings.

At the next session, after placing her into hypnosis (using only a few words based on a post-hypnotic suggestion I had given her in a prior session), she quickly went into a deep state.

I then had her become aware of her disturbing feelings, caused the feelings to become more and more intense, and then placed her on her traveling cloud and told her she would be going back to an earlier time and place that had to do with these very same feelings. This is a hypnotic technique commonly called "affect bridging."

The following transpired:

RTJ: Now we're still on our cloud. You're going to be able to go down, without any pain or discomfort. Now tell me Carol, where are you?

C: I can't tell. It's a long time ago. The walls...they're rock.

RTJ: They're walls...and rocks? Was this in a time before the present lifetime?

C: Yes.

Carol is quite agitated here. She obviously is in a very deep state of hypnosis and is struggling to talk. I am reluctant to lighten her state of hypnosis as I feel it important to stay accessing her subconscious in this deep state.

RTJ: Now look around and tell me...what's occurring?

C: It's like volcano rock...and.

RTJ: Tell me what you're feeling and thinking.

C: There's no way out. There's lots of people.

RTJ: Tell me about yourself. How old are you?

C: Twenty.

RTJ: Are you a man or are you a woman?

C: I'm a woman.

RTJ: Look around and tell me, what do you see?

C: There's...outside the walls there's trees.

RTJ: Are there people there that you know?

C: (With great emotion.) Oh yes.

RTJ: Look down at your arms. What's the color of your skin?

C: Brown.

RTJ: What's your name?

C: (Carol is leaning forward, struggling to talk and showing great emotion.)

	Something like...My...Mya...I don't know.
RTJ:	Now there's somebody you know standing right next to you. And they're talking to you. What language are they using?
C:	Something like a Polynesian.
RTJ:	Now this person right next to you is talking to you and calling you by your name. What name are they calling you by?
C:	Mia...I think.
RTJ:	Why are you feeling so bad? What's going on?
C:	(Carol is crying now.) We're all going to die.
RTJ:	How do you know that you're all going to die?
C:	(With desperation.) They won't let us out.
RTJ:	Can you see who your guards are? Can you see who's holding you there?
C:	They're on the wall.
RTJ:	Do they have weapons?
C:	They're spears.
RTJ:	How are they dressed? What kind of clothes do they have on?
C:	I...I think...just like...they're just clothes you know...like a loin cloth.
RTJ:	Do they have anything on their feet that you can see?
C:	Yes. They have on sandals.
RTJ:	Now look down at yourself. Are you wearing anything?
C:	I don't think so.
RTJ:	Are you wearing anything on your feet?
C:	No.

RTJ: Now look at the other people that are there. Are they wearing clothes?

C: (Crying.) No.

A little later.

RTJ: Why are they holding you prisoner? Why are they holding you in this place?

C: They want the land.

RTJ: What kind of land is this?

C: It's good for farming.

RTJ: Do you have a mate?

C: I don't know…but there's a man there with me.

RTJ: And what's his name? Look at him. He's standing right next to you.

C: Chia (Phonetic. Then she adds:) That's not quite right.

RTJ: Look at his face. Look at his eyes. Are his eyes round or oriental looking?

C: They're round.

At this point Carol is so distraught I decided to leave her place of imprisonment, even though I had a lot more questions to ask.

RTJ: Now I want you to go forward to the time you died in that lifetime. You won't feel any pain or discomfort. Tell me, how old are you now?

C: Sixties.

RTJ: You're in your sixties. Can you see yourself? Where are you?

C: On a pallet in a house.

RTJ: Is there anybody there with you?

C: Yes.

RTJ: Tell me, who's there with you?

C:	My grandchildren.
RTJ:	Oh that's great. Tell me, what are some of their names?
C:	Mea…Tinka…Wan.
RTJ:	You've got a lot of nice grandchildren.
C:	Yeah (Smiling)…they're good.
RTJ:	Now you'll feel no pain or upset. After you've died. Can you see yourself now?
C:	Yes.

Here we encounter another unusual method of disposing of a body.

RTJ:	Where are you that you can see yourself after you died?
C:	I'm lying on a wood structure.
RTJ:	What's happening to your body on this wood structure?
C:	It's floating away.
RTJ:	Is it in the water?
C:	Uh huh.
RTJ:	And where are you that you can see this?
C:	In the sky.
RTJ:	Now look down. Can you see where it was that you lived? Was it an island…or a big continent?
C:	It was an island.

At this point I decided that I should bring Carol back to the present time. I had kept her in hypnosis for a prolonged period and she appeared exhausted. I put her back up on her cloud and started bringing her back to the present. However, on the cloud coming back to the present time, we had the most remarkable conversation.

RTJ:	Now I want you to flash back in time…flash back in time to the experience that you had where you were a prisoner. Tell me, in modern terms. What was the date that all that happened?

C:	It was before….it was before the Druids 'cause I saw them. I saw them coming and going…and I wanted to go back and make it right. I don't know when the Druids were.
RTJ:	But it was before the Druids?
C:	But there was something wrong. I wanted to go back and fix the Druids.
RTJ:	Was it the Druids that were holding you prisoner?
C:	I don't know because I don't know the Druids.

With Carol, as with all of my subjects, immediately after bringing her out of hypnosis, I asked about her experience in an attempt to get more detail. Usually the subjects saw many things I didn't know they had seen, and therefore didn't inquire about. You will recall Jane in Chapter three, who had regressed to being a black woman living in the jungle. Only after the session did she tell me she had also been a pygmy.

In talking about her experience, Carol said that the walls around her and the others were made of rocks. The people who were confined with her were from her town; friends. When I told her that she was apparently a Polynesian, she laughed and said that was a shock to her as she didn't know anything about the Polynesians.

Most astounding was that Carol told me in going back to the life on the island, and in coming back to the present, she saw flashes of other of her past lives in passing. She said both going and coming, the word "Druids" kept popping into her mind, and that in spite of my instructions to keep going, she almost stopped there.

Two matters of importance came out of the session that I felt needed to be explored further as a part of getting to the origins of her disruptive feelings.

First, that she was not killed in the confinement on the island was apparent. She lived to an old age, had children and grandchildren. Second, I felt it would be desirable to explore the lifetime she described as the "Druids".

Carol stated that she didn't know who the Druids were, or in what time period that they existed. I admitted my knowledge of the Druids was skimpy at best.

Carol agreed to another session to be held in the future. That's next.

10

Always Been Friends with Yager—We fight together

Carol was busy. She had a job, a family, and was engaged in other activities. I was busy continuing with my research, trying to hold one session a day to reach my goal in my first research project of at least 100 sessions with different subjects. Five months later Carol and I finally held our next session. I had reviewed the tape of our last session and was looking forward to working with her again.

After inducing hypnosis, deepening, and placing Carol on her traveling cloud, I again took her back to her life as Mia, living on the island. I was primarily interested in how she escaped from confinement, although I also wanted to explore her life-style as Mia.

When she signaled me that she was there, I brought her down from her cloud. She was Mia again, wearing clothes now ("Just a piece of cloth wrapped around"), with a mate named Tia.

The following then occurred:

RTJ:	Now you're standing right next to your mate, and you're talking to him. What are you saying to him?
C:	About the fields.
RTJ:	What are you telling him about the fields?
C:	We have to hurry.
RTJ:	Why do you have to hurry?
C:	Cause they're coming…lots of men…in boats.
RTJ:	All right. Now I want you to go forward to the next significant event. Tell me, what's happening now?

C:	They've got us all bunched up. In a little valley with rocks all around. Very frightened.
RTJ:	Are you wearing any clothes?
C:	No.
RTJ:	Look around. Are others wearing clothes?
C:	Not very many.
RTJ:	Again, now, go forward to the next significant event. What's happening now?
C:	We're going to get out.
RTJ:	How are you going to get out?
C:	Have it figured out. There's a crack in the wall.
RTJ:	Did you get through the wall?
C:	Yes. Not everyone is going out with us.
RTJ:	What happened to the ones that didn't get out?
C:	They died.
RTJ:	How did they die?
C:	They got killed.
RTJ:	Now go forward in time to another significant event. Where are you now?
C:	The enemy is gone now.
RTJ:	About how old are you now?
C:	Thirty-two.
RTJ:	Do you have any children?
C:	Four.
RTJ:	What's the name of the oldest?
C:	Tagen. (Phonetic.)
RTJ:	Now he's standing right next to you. You're talking to him. Tell me, what are you saying to him?

C:	Now you have to be a good leader.
RTJ:	Now you can hear yourself. Tell me in the same language you are speaking in. What are you saying to Tagen?
C:	(After a long pause.) I can't say it.
RTJ:	Now I want you to see yourself. You're sitting down at a meal. Tell me, what are you eating?
C:	Kind of like a rice...other stuff in it.
RTJ:	What are you eating with?
C:	Fingers.
RTJ:	Look around. Is there anyone else there with you?
C:	Yes. It's a celebration.
RTJ:	What kind of a celebration is this?
C:	It's about Tagen.
RTJ:	What's happening to Tagen that you're celebrating?
C:	It's a step for him to become a leader.
RTJ:	Are there any stone statues or religious objects around the island?
C:	Yeah...out of the rocks.
RTJ:	Look at one of the statues. What does it look like? Is it very big?
C:	Oh yeah. It's bigger...it's taller than I am. Its hair stands up. Has eyes big ears. It's mostly head.
RTJ:	Is there more than one statue?
C:	I don't see any.
RTJ:	Where is this large statue located?
C:	In our village.
RTJ:	Are there other statues on the island?

C:	Only see one.

When Carol described the statue as being "…mostly head", I had visions of the statues found on Easter Island. Carol later gave me a drawing of the statue, which doesn't look much like the Easter Island statues. Carol's drawing is shown at the end of this Chapter.

During this session as Mia, Carol was again in a deep state of hypnosis, but was not agitated or disturbed as she was in her prior session. Apparently we had cleared the emotion arising out of that experience. I felt it was time to leave Mia and go on.

I placed Carol back on her traveling cloud and took her to the lifetime she knew as the Druids.

She signaled me that she was there.

RTJ:	Now come down from your cloud…without any fear or discomfort. Tell me, how old are you now?
C:	I'm very young.
RTJ:	Now I want you to go forward in time. Go forward until you're about fifteen-years old. Tell me, are you a boy or a girl?
C:	I'm a boy.
RTJ:	Look down at your arms. Tell me, what's the color of your skin?
C:	It's light…but tanned.
RTJ:	While you're looking down, tell me, are you wearing any clothes?
C:	Feeling heavy.
RTJ:	Feeling heavy. What are the clothes made of?
C:	Rough. All brown. Some kind…flax or something.
RTJ:	Look down at your feet. Are you wearing anything on your feet?
C:	Uh huh.
RTJ:	What are you wearing on your feet?

C: They're skin boots…but they're soft.

RTJ: What's your name?

C: Yon.

RTJ: Are your mother and father still alive?

C: I think so.

RTJ: What's your father's name, Yon?

C: Dirk.

RTJ: And what's your mother's name?

C: Mary.

RTJ: Do you have any brothers and sisters?

C: Uh huh.

RTJ: How many brothers and sisters do you have?

C: Three.

RTJ: Are they girls or boys?

C: Two girls and a boy.

RTJ: What's the boy's name?

C: Erik.

RTJ: What are the two girls' names?

C: Kathleen…(Long pause.)

RTJ: Don't worry about it. You can tell me later. Now look around and tell me, Yon, where are you?

C: The sky's dark…but it's day.

RTJ: Do you see any dwellings of any kind?

C: Yes. It's a town.

RTJ: Look at one of the dwellings. What's it made of?

C: Rocks…and straw.

RTJ:	I want you to walk up close to one of the dwellings. And tell me, how big is this dwelling?
C:	Not very big.
RTJ:	Is it more than one room?
C:	No. It's round.
RTJ:	What kind of a floor does it have?
C:	Dirt.
RTJ:	What kind of a door does it have?
C:	There are rocks around…and skin for the door.
RTJ:	Does this town have a name? What do you call this town?
C:	Yorg. (Or possibly York.)
RTJ:	Do you have a leader?
C:	Yes.
RTJ:	What's his name?
C:	Ian. (Phonetic.)
RTJ:	Now look around. What do you see on the terrain? Do you see oceans or mountains?
C:	Mist…and marshes…and rolling hills.
RTJ:	Now I want you to go forward in time. Go forward until you're about twenty years old. Look around. Tell me, where are you now?
C:	Close to the black rocks.
RTJ:	What are you doing there?
C:	Praying.
RTJ:	Who are you praying to?
C:	To God.
RTJ:	What do you call this God?

C:	(With some impatience) Everyone knows his name.
RTJ:	What's this God called?
C:	Yadrick. (Phonetic.)
RTJ:	Do you have more than one God that you pray to?
C:	The others do, but I don't.
RTJ:	Is there a special dwelling where you go to pray?
C:	It's a place.
RTJ:	What do you do for a living?
C:	We go in boats.
RTJ:	What do you do in the boats?
C:	We go places in them. Sometimes we trade. Sometimes we just take things.
RTJ:	Do you have any weapons with you?
C:	Spears and swords. and like a hammer.
RTJ:	What do your people call themselves?
C:	Tribaveen. (Phonetic.)
RTJ:	Now Yon, you're on your boat and there's someone right next to you, and they're talking to you. What are they telling you?
C:	We're going to the village to get stuff…take it.
RTJ:	Now listen very carefully. He's talking to you about going to the village and taking all this stuff. Tell me what's he saying…in the same language.

Here Carol speaks a number of words in what sounds like a foreign language. It's a guttural sounding language.

RTJ:	Now I want you to go forward in time to the next significant event. What's the next significant event, Yon?

C:	(With emotion.) They're going to kill the people. (Carol is crying now and is very agitated.) I tried to stop them...They wouldn't listen.
RTJ:	Now, go right past that event. Go forward in time...go forward in time to the next significant event. Now tell me, what's happening now?
C:	(Carol is still very emotional.) I'm really mad...I told my God that too.
RTJ:	All right. Now go forward in time...forward in time to the next significant event. Where are you now?
C:	I'm back with my people.
RTJ:	Look around. Where are you?
C:	In the house. It's not the same house, but it's my house.
RTJ:	How old are you now?
C:	I'm twenty-four now.
RTJ:	What are you doing?
C:	(Carol is speaking in a very agitated manner now) I'm with the people I love...family. I'm trying to stop people who kill. They were just people. I'm trying to stop them. They're my people too, but they're different. They're not the same. We're talking about it. We're planning. How are we going to stop them?
RTJ:	How are you going to stop them?
C:	We'll give them wrong directions. We'll warn the people that they're coming.
RTJ:	Now I want you to go forward in time...until you're about forty-five years old.
C:	I'm forty-three.

Notice here she doesn't follow my suggestion, but corrects me.

RTJ:	All right. Do you have a mate now?

C: Yes.

RTJ: What's her name?

C: Katrina.

RTJ: Do you and Katrina have any children?

C: Yeah, two.

RTJ: Are they boys or girls?

C: A boy, Yelsa...and a girl.

RTJ: Now Yon, I want you to see yourself, sitting down at a meal. What are you eating?

C: A bird.

RTJ: How is the bird being fixed?

C: Oh, it's being roasted.

RTJ: What kind of a bird is this?

C: It's...like a turkey bird.

RTJ: How did you roast this bird?

C: Over the fire.

RTJ: Are you eating anything else besides the bird?

C: Soup...clear soup...bread. There's something out of the marshes...a plant.

RTJ: Look around...tell me, where are you that you're eating this meal?

C: At home.

RTJ: Are there tables or furniture?

C: There's a table...we're sitting on the floor near a small table.

RTJ: What is the soup contained in?

C: Metal bowl...round.

RTJ: Are there any kind of utensils that you're eating with?

C:	A knife…a two-pronged fork.
RTJ:	What's the fork made of?
C:	(With impatience.) Oh…everything's metal.
RTJ:	Look at your mate. Is she wearing any kind of jewelry?
C:	It's just leather.
RTJ:	What's she wearing?
C:	A shift. Just a shift.
RTJ:	Look outside. What's the weather like?
C:	It's real misty. It's cool. Moist.
RTJ:	Is there snow there?
C:	No snow right now. Lots of snow later.
RTJ:	Now Yon, I want you to go forward in time to the day that you died. Without any pain or fear. Tell me, how old are you now, Yon?
C:	fifty-four.
RTJ:	Look around. Where are you?
C:	In the boat.
RTJ:	Why are you dying?
C:	I've been stabbed with a sword.
RTJ:	Why were you stabbed with a sword?
C:	We were fighting.
RTJ:	Who were you fighting?
C:	Those people again.
RTJ:	Are they part of your tribe…these people that you're fighting?
C:	They're the same people.

RTJ:	Now I want you to pass right through that death experience...without any pain or fear. Now you've just died. Can you see your body?
C:	Yeah.
RTJ:	Where are you that you can see your body?
C:	Well...I'm just there looking...up above. I can see it there in the boat.
RTJ:	Is there anybody up there with you?
C:	Yeah.
RTJ:	Who's up there with you?
C:	There's others.
RTJ:	How do you know they're there. Can you see them?
C:	I can feel them. I can talk to them.
RTJ:	Look around. Who's there that you know?
C:	My friends. Those who died in the war.
RTJ:	Where are you going now?
C:	To see that the family are all okay...taken care of. Then join the others who are far away.
RTJ:	Now I want you to go forward in time. You still haven't been born into another lifetime. Look around and tell me. Where are you now?
C:	It's softer now.
RTJ:	Can you see anything?
C:	Kind of a gray mist.
RTJ:	Have you received any instruction during this period of time between lives?
C:	I understand it better now.
RTJ:	Has anybody told this to you...helped you arrive at this understanding?

C:	I don't know how I got it…(Long pause…then with surprise) Yeah.
RTJ:	Have you received any punishment during this period between lifetimes?
C:	No. No punishment.
RTJ:	Look back at this lifetime where you lived as Yon. Look back at your family and your friends. Did you know any of those people in other lifetimes?
C:	Always been friends with Yager. We fight together.
RTJ:	Do you know any of those people in your present lifetime?
C:	Sometimes you know them.

I had had Carol in hypnosis for a long period of time so I brought her back to the present. In the post-hypnotic interview, she told me some fascinating things.

When she was Mia, she said other statues were around the island, but only one in her village.

I asked her to describe the boats she saw when she was Yon. She said they were long and thin, but not very deep. They were more square than a canoe, but much bigger, and they would hold as many as twenty people. She said they had both sails and oars. She also commented that, "It's strange, I don't see any seats."

She also said, "When I was hurt, the sea was really rough. Everyone was upset and concentrating on me. Others were hurt too, but I was the only one that died. I wanted to get back home to die, but I died right there in the boat. I felt like I was a leader. Not like a king leader."

I asked her if this was the life where you "fixed" the Druids. She answered, "Yes, what ever a Druid is."

She described the rocks where she prayed. They were black because of fires. Not volcanic fires, but man-made fires, made to honor their Gods. She later drew me a picture of the rocks. It's attached at the end of this chapter.

If we do have past lives, what was it like the first time we lived on earth? Ellen seemed to have this experience as we will see in the next chapter.

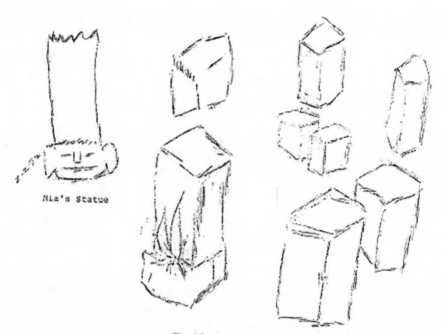

Mia's Statue

The black rocks where Yon prayed

11

Cro-Magnon...or Possibly Earlier

Ellen is a thirty-seven-year-old, white, married lady, with three children. She is a college graduate. On her Survey, she stated that her religious affiliation was Catholic, but she indicated no involvement in religious activities. She stated that she believed in some form of life after death, and listed "uncertain" as to whether or not she believed in reincarnation. Her self-report hypnotic depth was five.

Her first past life was as a young, white girl named Alisa, and when I joined her when she was in hypnosis and had regressed, she was standing near a mud hut. She resisted my attempts to get her to go inside the hut, but I finally talked her into going to the doorway and looking in by promising that it would only be for a few moments. She didn't know who her mother or father were. Her existence in that lifetime was very depressing for reasons I was never able to develop. Her resistance to my questions was very strong. She died at age sixteen "'cause I want to." After she died, and after I had explored her existence in-between lives, I tried to take her forward to another life.

RTJ:	Now I want you to go forward in time. You're just about ready to be born into another lifetime. Tell me, are you choosing to be reborn again?
E:	No. I don't want to go through that again.
RTJ:	Is anybody telling you that you have to be reborn again?
E:	Yes.
RTJ:	Why are you being reborn again, do you know?
E:	(With anguish.) No. I don't know (Ellen is crying now and seems quite agitated.)

RTJ:	Now let's go back up on our cloud again where it's safe and secure. Now the cloud is going back through time and space...Back before you lived as Alisa. Back before that lifetime. Back until you see yourself in another life-time. When you see yourself there, let me know by rais-ing your right index finger.

Although Ellen was a good subject, in a medium stage of hypnosis, she refused to go into another lifetime. Eventually I brought her back to the present time.

In the post-hypnotic interview, she said, "God Bob, it was so...just despair and hate." (She was still sobbing.) "I really felt everything that she felt. I wanted to die. I didn't want to live anymore. It was so depressing. It was totally beyond hope."

Ellen was an excellent example of revivification. When I asked her if she thought her experience as Alisa was real or imagined, she answered, "I was physi-cally in that body, looking out of those eyes."

Although her first experience was quite unpleasant, she asked that I work with her once more, so she could experience another past life.

Our second session was only five days later. After induction and deepening, I took her back "...to the first time you experienced life on this earth." I hadn't told Ellen that that was what I planned to do.

RTJ:	Now without any pain, fear or discomfort, come down from your cloud.
	Look around and tell me, what do you see?
E:	I'm in a cave.
RTJ:	How old are you?
E:	I'm about ten.
RTJ:	Are you a boy or are you a girl?
E:	I'm a boy.
RTJ:	Now look down at your arms. What color is your skin?
E:	Kind of a...a little dark brown, but they tease me because I'm not as dark as everybody else.

Notice the volunteered information.

RTJ:	Look down at yourself. Are you wearing any clothes?

E:	Just skins…not much.
RTJ:	Are you wearing anything on your feet?
E:	Something. It's just a bare sole kind of thing. It's not much.
RTJ:	Is there anyone there in the cave with you?
E:	Yes, there's a man. He's walking around the fire.
RTJ:	Who is this man? Do you know him?
E:	I think he may be my father. But we don't have…don't have fathers like you would think.
RTJ:	Now I want you to walk outside the cave. Look around. What do you see?
E:	It's not a deep cave. Just sort of a hollow out. Kind of on a hill.
RTJ:	Look around. What do you see?
E:	Just kind of grass lands.
RTJ:	Are there hills and mountains?
E:	No. Hills way far away.
RTJ:	Now I want you to go forward in time…until you're about fifteen years old. Tell me, where are you now? Are you still around this cave?
E:	We're hunting. We're out in the grasslands.
RTJ:	What kind of weapons are you carrying?
E:	Spears.
RTJ:	What kind of animals are you hunting?
E:	I've never seen them before. I don't know what they are.
RTJ:	Look real close at them now. Describe them to me. What do they look like?

E: Some kind of antelope but they don't have big horns...real small horns.

RTJ: How big are the animals?

E: They're pretty big. About the size of a cow or something. A little bigger.

RTJ: Look around and tell me. Who's there with you on this hunting trip?

E: There's people all scattered around...yelling.

RTJ: Now I want you to go forward in time. Go forward until you're about 25. What's your name?

E: We don't have names.

RTJ: How do you communicate with each other? Do you talk to each other?

E: Not words. Just mostly gestures.

RTJ: Do you have a mate now?

E: Yes.

RTJ: Do you have any children?

E: I don't see any. I'm back in the cave. I don't see any.

RTJ: Now I want you to see yourself eating some kind of meal. What are you eating?

E: The...the meat. It's white. Looks like pork almost.

RTJ: Do you cook the meat?

E: Yes.

RTJ: What do you eat the meat with?

E: Our hands. I'm holding it in my hands.

RTJ: Now I want you to go forward to some kind of a ceremony. Tell me what the purpose of this ceremony?

E: I'm back when I was getting a mate.

RTJ: What happened during this ceremony?

E:	We're around a fire...and there's a small group.
RTJ:	Is there anybody officiating at this ceremony?
E:	No. We're just standing in front of the fire. People are saying things.
RTJ:	Did you choose this mate?
E:	Yes.
RTJ:	Did she agree to be your mate?
E:	She didn't have a choice.
RTJ:	Look at her very closely now. Describe her to me. What color is her hair?
E:	Very dark. Not black. Very dark eyes. Very dark skin. Very round face.
RTJ:	Do you like this girl?
E:	She's okay. I know she'll give me lots of children.
RTJ:	Now I want you to go forward to the next significant event. You've just chosen a wife.
E:	We just had a son die.
RTJ:	How did he die?
E:	We don't know. He's very young. He isn't walking yet.
RTJ:	Go forward to the next significant event. What's happening now?
E:	I think I'm dying.
RTJ:	Do you know how old you are?
E:	I'm not even thirty yet.
RTJ:	Why are you dying?
E:	Something internal. It's not right.
RTJ:	Now without any pain or fear. Go right through that death experience. Now you've just died. Can you see your body?

E: Yeah...yeah.

RTJ: Where are you that you can see the body?

Here we encounter a very primitive form of body disposal.

E: I'm on some...like a constructed...boughs and trees...constructed limbs. Just laid out there.

RTJ: Where are you that you can see this body?

E: Outside somewhere.

RTJ: What happens to this body? Is it just left there?

E: Yeah.

RTJ: Now I want you to go forward in time. You haven't been born into another lifetime yet. Look around and tell me, where are you now?

E: A lot of nothingness.

RTJ: Is anybody there with you?

E: No. I hear some voices...but there's nobody there.

RTJ: Now I want you to go forward in time. You're about ready to be born into another lifetime. Tell me now, are you choosing to be born again?

E: Yeah.

RTJ: Is anybody helping you choose?

E: No.

RTJ: Are you choosing what your sex will be in the next lifetime?

E: No. I don't think I have a choice.

RTJ: Are you choosing your new parents in your next lifetime?

E: Not specific parents.

RTJ: Why are you being reborn?

E: Anxious to try it again.

Again, Ellen was a subject who regressed to being of a different sex and race. The culture in which Ellen lived as a boy was so primitive, that I wondered if it were possible that I might be the first Homo Sapiens to have a conversation with a Cro-Magnon, or some earlier species.

In any event, it now seems apparent that that part of our individuality that survives our physical death, existed at a very early time in our evolution, and very strongly indicates that we are spiritual beings, having continuous experiences in different life-forms.

Aside from the phenomenon of Ellen regressing to such a primitive culture, here is an example of only a partial dissociation while in hypnosis that I found interesting. You will recall in Chapter five that I mentioned that in hypnosis, a dissociation occurs—a partial separation between the conscious and subconscious minds allowing the hypnotist to interact with the subconscious mind without interference from the conscious mind.

In Chapter seven, with Jane in her past-life personality of Inkwa, she didn't know what a pig was, even though in her next past-life as an oriental man, and in her present life, she did know about pigs. The reason for her inability to recognize pigs while as Inkwa, was because Jane was in a deep state of hypnosis and the conscious mind had been pushed aside. The dissociation there was very strong.

Here, with Ellen, she is only in a medium state of hypnosis, the dissociation is not quite as strong. Ellen is able to partially access her conscious mind, even though she is reliving the personality of the very primitive, nameless male. She is able to describe the animals she saw as "Some kind of antelope...", and "About the size of a cow...", and "The...meat. It's white. Looks like pork almost". These comments, and others that she made, are references to knowledge contained in her present, conscious awareness.

Several days after the session, Ellen told me she had been so intrigued by her experience, that she had researched prehistoric animals at the library. She found drawings that looked very much like the animals she saw in her regression. The animals she saw had humps on their withers, but did not have large antlers. Ellen furnished me photocopies of the ancient animals that she described and what she saw was possibly a species of the ancient Red Deer. That species flourished in the Middle and Upper Pleistocene Epoch, possibly more than 35,000 years ago.[19]

In the post-hypnotic interview, Ellen told me that her group was hunting the females of the deer-like animals as they were less dangerous than the males. She said they would crowd them together in a small group before the final attack. They were hunting in high grass and rolling hills with natural caves in them. The caves were just hollows-out of hill sides, not long "tunnel" caves. She also said that the culture to which she had regressed had more of a social structure than what she had indicated when she was reliving that personality.

Importantly, Ellen said that before she researched the matter, she had never seen drawings or pictures of prehistoric deer or elk, and didn't know that they had existed.

In Chapter Twelve I want to discuss the past-life research of other serious researchers with the aid of what we have learned from the past lives accessed in my First Research Project.

After Ellen's experience, in my Second Repesearch Project, along with other events occurring in our past lives, I do attempt to explore our past lives the first time we lived on Earth, *in any form.*

12

What Are We to Make of It?

I selected the past-life regressions that I've set forth in the preceding chapters and those in the following chapters, from the experiences of my own subjects. I wanted to demonstrate that hypnotic regressions with healthy, normal, adults are not the freakish hallucinations of mentally troubled persons. I also wanted to demonstrate the very subjective nature of hypnotic regressions.

I believe this approach will aid the reader in evaluating the past-life phenomena, particularly those experiences of other researchers that have been validated.

Almost all of the past-life personages to whom my subjects regressed were not those of prominent persons, but were very ordinary. Those regressed lifetimes were usually to those who were poor, with the usual family and survival problems that you might expect for people living in those time-periods. These results were similar to the experiences of others working with past-life regressions.

One lady remarked after I brought her out of the hypnotic trance, "That was stupid." I asked her what the problem was, and she answered, "Why couldn't I have been someone important?" She regressed to being a ten-year-year-old boy, living on the side of a mountain herding goats.

Possibly a Cleopatra or a Napoleon are out there somewhere. I just haven't encountered them

Some things resulting from my First Research Project do seem quite evident, are not obvious from the research of others, and I believe to be quite important.

As disclosed by each of my subjects' Information Surveys, none was in therapy or suffering from a serious physical illness, and none seemed psychologically disturbed.

Most of my 104 subjects (seventy-eight per cent) did regress while in hypnosis to what seemed to be past lives, indicating that such phenomena is common among normal, healthy adults. It's an area of human experience like any other and is not a unique experience.

That regressions to past lives are the experiences of healthy, normal people adds to the importance of this phenomenon. It strongly indicates that past lives probably exist for all of us.

Looking at the educational backgrounds of my subjects, it would appear that an interest in the survival of our physical deaths and rebirth is compatible with intellectual competence. Almost eighty per cent of the subjects in my First Research Project had some college. Approximately thirty-nine percent were college graduates. Three had Masters Degrees and two were PhDs. The question of whether or not life exists beyond physical death seems to intrigue a broad range of our population. A summary of the composition of my subjects in my first Research Project appears in Appendix B for those who may be interested.

I found it important and very interesting that an analysis of the data from my First Research Project using a computer statistical analysis program, indicates that a person's religious beliefs, religious involvement, belief in life after death, education, and whether they expected to recall past lives had no effect on whether or not they regressed beyond birth while in hypnosis, and made contact with what seemed to be a past life. (Again; for those interested, the analysis appears in Appendix B.)

A number of criticisms concerning the use of hypnosis as an investigative tool into the past-life phenomenon, other than those mentioned in prior chapters need to be considered to evaluate my research and the research of others. These criticisms seem basically to fall within the categories of fraud and deceit, fantasies based on information to which the subjects were once exposed but have forgotten (cryptomnesia), genetic memory, extrasensory perception, the effects of a hypnotic trance suggestibility, spirit possession, and that gaps in memories can be filled-in with memories that are false but that the subject accepts as true (confabulation).

When you consider the large number of apparently emotionally stable, mature subject with whom I worked, who sought no personal gain from the research, and who will remain anonymous, fraud seems extremely unlikely. No financial consideration was paid to or received from any of the subjects. No motivation or opportunity for either fraud or intentional deception surfaced.

Cryptomnesia is the concept that the regressed subjects had somehow come into contact with the information, persons, and/or incidents given in their past lives, and had then forgotten the source, and reported it in the regression as though it were an actual experience. Presumably most of my subjects had read many books, attended movies, watched television, traveled, and engaged in other activities that would give them a great deal of background information that some critics believe is the basis of their regressions.

But of my 104 subjects, *twenty-three did not regress beyond birth*. Presumably these twenty-three non-regressors had also read many books (seventy-three percent had attended college), watched television, traveled, been to movies, and engaged in other activities, that would provide them with a great deal of background information.

Thirteen of these twenty-three had reached at least a medium-deep state of hypnosis. If cryptomnesia were a significant factor in the past-life phenomenon, then these non-regressors would most likely have regressed and reported past lives with this once-learned-but-forgotten information as did the others, if in fact the others did so.

Cryptomnesia hardly seems borne out by the subjects' mundane experiences and I wasn't personally convinced that this was an important element in my subjects' experiences.

The remote possibility of cryptomnesia, however slight, was one of the reasons I did not attempt to take the subjects back to specific time-periods or places as this could conceivably give rise to speculation by the subjects about what those times and places would be like.

In post-hypnotic interviews, very few indicated any knowledge or interest in the time periods or places that appeared in their past lives. If indeed some information had been obtained and then forgotten, the past-life regressions would have likely acted as a cue, helping them remember obtaining some past information similar to that brought out in their past lives.

In several cases, before their hypnotic sessions, the subjects indicated a special interest in a particular place or time period. In no instance did such subjects regress to those time periods. For example, one subject who had been in Russia after World War II, stated he had had *deja vu* experiences while there, as if having been there before, when he had not. However, neither of his two past-life experiences was in Russia.

Genetic memory is advanced by some as a possible explanation of the past-life phenomenon. This is the concept that somehow, through the genes we inherited from our ancestors; the memories of our predecessors' lives survived and were passed down to us.

Genetic memories, if they exist, require the unbroken passage of genes through our ancestors and then down to us. My subjects remembered surviving their deaths, floating out of their bodies into another, non-earthly realm, reciting incidents that happened in this other dimension before being reborn again. Obviously, no genetic material could have been passed down of events that occurred after the death of their bodies. Deceased entities don't reproduce.

If extrasensory perception (ESP) is at work here, as claimed by some, it must be available on demand and on a very large scale, and be most prevalent while only in hypnosis. Most of my subjects did not exhibit ESP abilities outside of the hypnotic regressions, which seems to negate their use of ESP while in hypnosis.

An additional problem with the ESP concept is apparent in cases of xenoglossy, where the subjects spoke responsively in a foreign language not learned in their present lives. Possibly, knowledge and information can somehow be transferred from the minds of one living person to that of another (ESP), but not knowledge of how to do a skill, such as swimming, dancing, bicycle riding and speaking responsively in a foreign language not learned in their present lives. These skills must be learned by practice, not by ESP.

To say that each of my subjects who regressed received the information concerning these past lives by ESP, transferred from presumably the minds of persons now living, and then to identify with the deceased personalities in their regressed past lives, is to extend the ESP doctrine to such lengths as to be beyond credible consideration.

I discounted spirit possession as a viable explanation of my subjects' experiences. It seemed highly improbable that each of my eighty-one subjects was possessed by two different spirits each session and only when the subjects were in hypnosis, and not in their normal lives while not in hypnosis.

Additionally, each subject, with two exceptions, identified with the regressed personality and reported their experiences in the first person. As Ellen (Chapter eleven) reported, "I was physically in that body, looking out of those eyes."

One claimed explanation of the past-life phenomenon while a subject is in hypnosis is that the subject is just responding to and complying with the suggestions of the hypnotist.

After putting my subjects into hypnosis and deepening the trance to reach at least a medium-deep state, I suggested that they would go traveling on a cloud. I took them back to the age of fifteen, and then to age five, in their present lives. This acts as a form of verification for them. They are able to see incidents and persons they know that they encountered at those ages. Then I tell them that the cloud is taking them back before birth, to an existence before this lifetime.

Everyone is suggestible to some degree and as a generality, a person in hypnosis is more suggestible than while in a normal waking-state. The observation, at least in my research, that the hypnotist suggests to the subjects that they go back into time before birth, is correct. Generally the suggestions end there. No suggestions are made as to their sex, names, family relationships, events that occurred, and their methods of death. This information came from the subjects' subconscious minds.

As explained in Chapter six, when a subject is in a medium-deep state of hypnosis, dissociation occurs; that is, the conscious and subconscious minds are divided and can then be utilized as interdependent-yet-independent entities. In such a state, generally an increased responsiveness to suggestion occurs. This responsiveness however, is not to be confused with gullibility or non-critical acceptance of suggestions. You will recall in several of the dialogues examples of the subject's ignoring my suggestions appeared, usually because of events occurring in the regressed past life that were more important to the subject than my rather benign questions.

To attempt to reduce past-life regressions to merely a compliance with the hypnotist's suggestions would be difficult. If this were so, why didn't more of the twenty-three non-regressors, who were in a medium-deep state of hypnosis, also regress?

Some critics suggest that the subjects, while in hypnosis and while in a suggestible state, are trying to reinforce the hypnotists' beliefs in reincarnation. But at the time of my first research, I was neither a believer nor a disbeliever in reincarnation nor of the survival of our physical deaths and rebirth, and this was communicated to my subjects before the hypnotic sessions.

Confabulation can occur. Confabulation is generally defined as the filling-in of gaps in actual memories with memories that are false, but which the subject accepts as correct. Confabulation can occur both in and out of hypnosis, without any deliberate attempt to mislead or deceive. Any experienced police officer or attorney can confirm that this happens.

I cannot assure the reader that no errors arose in the subjects' memories of past lives, or that some form of confabulation did not occur.

Errors in memories occur to all of us in non-hypnotic states, and to assume that they don't also occur to a person when in hypnosis would be fallacious. Memories are not stored with the exactness of a voice on a tape recorder. Memories are stored on the basis of perception, and are subject to the same distortions as other perceptions. Hypnosis won't generally improve the original perception of what was encountered. What is being recalled is only how the person experienced it at that time.

I would add one further thing concerning errors. Dissociation in hypnosis seems to occur in direct relationship to the depth of the trance. However, subjects do fluctuate in depth during the trance, going both lighter and deeper.

When a subject was in what seemed to be a past life, if it was appropriate, I would frequently ask the year. Obviously to ask this of the lady who regressed to being the illiterate ten-year-old boy living on the side of a mountain would be of no use.

In those cases where I did ask for the year, I sometimes would get a direct response, with no hesitation. In those cases where the response was substantially delayed, I felt that the subject would lighten their trance state, and make what seemed to be a logical guess, seemingly accessing their conscious minds. As a result, I am very skeptical as to dates given in those circumstances. This could be called confabulation, but I attribute it more to a lack of depth with its concomitant lack of dissociation.

The question basically comes down to, "*Can* memory be distorted or be in error while a subject is in hypnosis?" and I would answer "Yes." But, if so, "*Must* memory necessarily be distorted or in error while a subject is in hypnosis?" Based upon my own experience and the experience of others, I would answer with certainty, "No."

The most vociferous critics of the possibility of our being able to contact past lives while in hypnosis seem to start from the premise that, "It can't happen, therefore it didn't happen". They then conclude that the phenomenon is caused by fraud, deceit, or some other explanation.

Many of the academic critics of past-life research either view the whole subject with scorn ("It just isn't science"), or they are advocates of a particular school or tradition in psychology (biological, behavioral, analytic, etc.) and will then try to analyze, interpret, and fit the past-life experience into the constructs of their particular discipline, rather than taking the experiences on their own terms. In discussing confirmed cases with many of these critics, I find that the cases are usually not rebutted, but simply ignored.

We humans are alone, among all living creatures, that know we will someday die. But what happens when death occurs? Considering my research, along with that performed by others, are we any closer to knowing if we survive our physical deaths to be born again?

Let's take a look at the research conducted by others in an attempt to put the matters into proper perspective. Much of this material is often not published in popular books or in magazines and journals readily available to the general public. However, if your local public library has an inter-library loan service, such material is easily obtainable, usually at no or little cost.

In my Introduction, I mentioned the works of two authors, Dr. Ian Stevenson, a Professor of psychiatry at the University of Virginia; and Dr. Helen

Wambach, a clinical psychologist (now deceased). These two have conducted studies into these phenomena that seem important to me.

Dr. Stevenson's research mainly involves investigating and authenticating past-life memories that arise spontaneously, not involving the use of hypnosis. He has investigated hundreds of cases where people, most often children, suddenly start recalling past lives. In many cases, he has been able to verify that people with the names, and located in the places and time-periods as recited by the subjects, have actually existed. Frequently these subjects can speak in languages other than that of their present culture (called xenoglossy). Usually as the child grows older and becomes more grounded in his/her present life, these memories fade.

An example of one of Stevenson's well-researched cases involving an adult where hypnosis was not involved, is in his 1984 book *Unlearned Language.*[20] He reports a case of an adult lady named Uttara Haddar, who was born March 14, 1941, in Nagpur, India. Her language was Marathi. The personality of a woman named *Sharada* emerged spontaneously for Uttara early in 1974 while she was in her thirties. Uttara's behavior changed and she developed a tendency to wander away saying she "wanted to go to a place where she thought she belonged." As Sharada, Uttara was unable to speak her native language, Marathi. Instead she spoke in Bengali. She also modified her dress and her conduct to conform to that of a married Bengali woman.

As Sharada, she preferred the foods of Bengal and had knowledge of these foods not expected of a woman in Uttara's region. She also exhibited a familiarity with the names and locations of small towns in Bengal. Uttara and her parents had no knowledge of the Bengali language or customs. The Sharada personality could speak Bengali fluently, and a substantial number of her statements about her life as Sharada have been verified.

Sharada gave many details including the names of her prior family in Bengal, and a family corresponding to Sharada's statements has been traced in the part of Bengal where she said she lived in the 1820s.

Two cases researched by Stevenson that did involve hypnosis, and which *do* indicate that hypnosis can be a valuable tool in exploring the questions of whether or not we survive our physical deaths to be reborn again, are the cases of *Jenson*[21] and *Gretchen.*[22]

The Jenson case arose in the middle 1950s. A woman, T.E., of Jewish parentage, was placed into hypnosis by her physician husband. T.E. was born and raised in Philadelphia.

Under hypnosis, T.E. regressed to being a man named Jenson Jacoby. The Jenson personality described his life as a peasant farmer in Scandinavia in great detail, and he spoke Swedish responsively; that is, he communicated sensibly in Swedish. The Jenson personality described his life in a tiny village—his crops, farm animals, food, family, and religion, which was Christian. He hunted bears, enjoyed playing games, drinking at the village tavern, and consorting with women other than his wife. He was killed by a blow to the head.

Stevenson estimated that Jenson lived in the Seventeenth Century although the specific village named by Jenson has not been located. Neither T.E. nor her husband or parents had ever been to any of the Scandinavian countries or knew intimately anyone who could speak any Scandinavian language. Stevenson's investigation into the backgrounds of the subject and her family, and the linguistic aspects of the case took some six years.

The Gretchen case arose in the 1970s in the Eastern United States and is also reported in Stevenson's book *Unlearned Language*. A Methodist minister hypnotized his wife Dolores, and gave her the suggestion that she should regress to "previous lives." While in hypnosis, she assumed the personality of Gretchen Gottlief, a young German girl living in the past in Eberwalde, Germany. Gretchen reported that she died at the age of sixteen.

Investigation revealed that none of Dolores' family or ancestors were German, and that she, Dolores, had no knowledge of the German language. Stevenson not only researched her background at great length, but also had Dolores take a polygraph test that affirmed her denials of knowledge of German before the regression. Gretchen not only spoke German, but spoke it responsively; that is, she gave sensible replies in German to questions put to her in that language.

In the case of Gretchen, Stevenson reported, "It has not been possible to trace any person whose life corresponded to Gretchen's statements." Xenoglossy seems to be one of the most compelling aspects of both the Jenson and Gretchen cases.

I was surprised that the people I knew who were curious or interested in the possibility of our having past lives were basically unaware of Stevenson's books and articles. I have listed many of these in the *Footnotes* at the end of this book.

Dr. Helen Wambach conducted her very extensive research by assembling a number of subjects in workshops. After putting them into hypnosis she regressed them back through five time periods—1850, 1700, 1500, A.D., and twenty-five and 500 B.C. She allowed each subject to chose one of those time periods to explore. She asked them specific questions about the lives they were living. After awakening, she had them fill out data sheets. She then used these data sheets in evaluating whether what had been reported was consistent with the clothing, architecture, climate, landscape, and similar matters for the time-periods and places to which the subject had regressed. Her work involved a very large number of subjects.[23]

I thought Dr. Wambach's research could be improved upon by working with subjects on a one-to-one basis, rather than with large numbers simultaneously. In a session with only one subject, I could more accurately determine that the subject was in a satisfactory state of hypnosis for regression. I could thus obtain the best past-life regression, if indeed the subject did regress. I also thought that not suggesting a specific place or time-period to the subject would be an advantage, thus avoiding any criticism that the hypnotist was influencing fantasy by his or her suggestions.

Also as I mentioned in my Introduction, there are two remarkable cases, substantially verified, by two therapists, Linda Tarazi [24] and Rick Brown.[25]

Tarazi's client, L.D., regressed while in hypnosis to the life of a Spanish woman named *Antonia,* born November 15, 1555. L.D.'s ancestry is German and her religion is Protestant. She is married with two children. In some thirty-six formal hypnotic-sessions, she described her romantic and adventurous life as Antonia, with great detail as to names, places, and dates. Tarazi spent over three years researching the life of Antonia, "…in two dozen libraries and universities, travel to Spain, North Africa, and the Caribbean, and correspondence with historians and archivists." She verified well over 100 facts stated by L.D. as Antonia, but uncovered no errors. Much of L.D.'s information could be located only in old, obscure Spanish sources, and some was found only in Spanish archives. Some of L.D.'s information given as Antonia correctly corrected the authorities.

Tarazi had used hypnotic regression in therapy sessions before the L.D. case arose, and she stated, "…nearly all of the 'previous personalities' evoked during these sessions are unverifiable and (she opined) almost certainly derive from fantasies on the part of the subject." After her experience with L.D., and after considering psychodynamic factors, fraud, cryptomnesia, role-playing, dissociation or multiple personality, genetic memory, racial memory, clairvoyance, precognition, retrocognition, telepathy, mediumship, possession, and reincarnation, she concurred with L.D. who accepted Reincarnation as the explanation of her experience.

A similar case that arose while the subject was in therapy, and which was also well verified, involved a client of hypnotherapist Rick Brown named Bruce Kelly. Kelly was born January 19, 1953. While in hypnosis, Kelly regressed to a past life as *James Edward Johnston.* Johnston was a crewmember on the U.S. Submarine *Shark.* A Japanese destroyer sank the Shark on February 11, 1942, and all persons aboard, including Johnston, were drowned. The life of the deceased Johnston as relived by Kelly while in hypnosis was verified by documents from the Civilian Conservation Corps, the United States Navy, and high-school attendance records. In addition several of Johnston's still-living friends and relatives substantiated information recalled by Kelly while in hypnosis.

There are many other cases where the past lives recalled have been verified as the lives of persons who have actually lived before. The foregoing cases of researchers Stevenson, Tarazi and Brown are only the tip of the iceberg, but do constitute "hard core" evidence that their subjects Haddar, T.E., Dolores, L.D. and Kelly have indeed lived before their present lives, in some form. I hope these cases and others will challenge those who are dogmatic skeptics or committed non-believers to investigate the past-life phenomena further.

My eighty-one subjects who regressed had at least two past lives, each life being in a different time-period and place than the other. Frequently a subject regressed to being of a different sex and race in each past life.

Considering the research of Dr.'s Stevenson and Wambach, together with my own, and the well-verified cases reported by Tarazi, Brown and others, the evidence for the survival of some part of our individuality after physical death, and for rebirth, can't be dismissed as a joke or as the aberrations of mentally-disturbed people. Obviously something is going on that can't be adequately explained by existing, generally held, physical or psychological concepts. And I would maintain that whatever that is, it's important and has meaning for our lives.

No significant body of thought presents strong empirical evidence that anything else is happening, other than that we do survive our physical deaths and are reborn. Stevenson's cases, where persons can responsively speak in languages other than those that they have learned, are especially persuasive. Xenoglossy is one of the most important evidentiary factors indicating the survival of a portion of the human personality after death and rebirth.

The skills involved in learning to speak a language are like the skills in learning to swim or ride a bicycle. You can't learn to speak a language without speaking it. You can't learn to swim or ride a bicycle without doing it. These types of skills are not communicable from one person to another by ESP, verbally, or otherwise. They must be learned.

If we have lived past lives, and the evidence from the "hard-core" cases is very strong, the implications are mind-boggling. For example, survival and rebirth may be unrecognized factors in personality development and in the resolution of psychological conflicts. Dr. Edith Fiore demonstrated this possibility in her book, *You Have Been Here Before*.[26]Past-life therapy under hypnosis can and has resolved and eliminated both psychological and physical symptoms as reported by Dr. Brian L. Weiss in his 1992 book *Through Time and Healing.*[27] The memories that surface from what seem to be past lives, impact therapeutically just as do memories from a person's present life.

For example, remember Carol's present-life's disturbed feelings about concentration camps and people dying (Chapter nine). In Carol's case, the regression back to what seemed to be her past life as prisoner on an island, knowing that she was going to be killed, together with the emotional clearing associated with that incident, acted as a catharsis. She is now free from those troubling feelings.

Survival and rebirth may be the explanation for homosexuality and sexual identity confusion, phobias, congenital deformities, and positively for the phenomenon of child prodigies such as Mendelssohn and Mozart, who both had advanced musical knowledge and skills at very early ages. Mendelssohn made his first public appearance as a pianist at the age of nine, and performed his first original compositions at eleven. Mozart composed five short piano pieces at the age of six that are still frequently played.

Stevenson suggests that past lives could be a unifying theory, tying together seemingly unrelated conditions from the fields of medicine, biology, psychology and psychiatry.[28]

So now, at the conclusion of my First Research Project, I repeat what I asked in my Introduction: "What are we to make of the almost universal human experience, that regardless of religious or other beliefs, while in hypnosis, we humans seem to have memories of our own past lives?" I concluded that *my* research (ignoring for the moment, the research of others) doesn't absolutely prove survival and rebirth, but it is certainly strongly consistent with those concepts.

However, the research of Stevenson, Tarazi, Brown, and others can't be ignored. These researchers are apparently of good character, well qualified, without collusion or interaction, but still verifying the details of their subjects' reported past lives. The obvious implications of their research are compelling and contribute new and substantive evidence of what it means to be human.

What was I to do next? I had arrived or been taken by the evidence to know that we are not just humans with a spirit that survives our death, but that we are spirits having a human experience.

Although it was gratifying that my article about my first research was published in one of the psychological Journals,[29] I didn't do the research with publication in mind. My research was a personal journey into the past-life phenomenon; a subject different from and even hostile to the mainstream culture into which I had been born and raised.

No attempt was made to obtain verifiable data in the regressions in my First Research Project, and practically none of those regressions are verifiable by means readily available. The skepticism that I had had about the possibility of the existence of past-lives led me to the obvious: why not conduct a Second Research Project to regress subjects to a time that would most likely give verifiable information? Assuming past lives to be real, this time would probably be the past life of each subject just before their present lifetime. If verifiable data could readily be obtained from subjects regressed before birth while in hypnosis, this would bolster the existing evidence that at least some of us do indeed have past lives, and give me a much firmer grasp on "What are we to make of it?"

In reviewing my first research, particularly the experience of Ellen (Chapter eleven) who I regressed to the first time she ever experienced life on Earth, and believing evolution to be a scientific fact, I also planned to attempt to take each subject back to the first time they ever lived on Earth *in any form*. I had no inkling of the questions this would open up for me, but I felt that a broader range of experiences would help me understand the past-life phenomena.

I thought I would also explore when the surviving spirit entered into the new mother's womb would be interesting. Was the fetus aware of its mother's feelings and emotions?

In my First Research Project, while bringing the hypnotized subjects back to the present on their traveling clouds, I frequently asked them to look up into the sky at all the stars and the planets, and tell me if they ever lived on another planet. Surprisingly, thirty-seven percent said that they had. What would the subjects encounter if I attempted to explore their experiences living on another planet at some time in the past? Again, I had no conception of what I would encounter, if anything. Some of the things I did encounter may possibly stretch the limits of your concepts of reality and what are "normal" human evolutionary experiences. It certainly stretched mine.

PART III
The Second Research Project

-to regress subjects while in hypnosis to their past lives just prior to this one in an attempt to get verifiable data;

-to regress subjects while in hypnosis to the first time they lived on Earth in any form;

-to explore when the spirit which survives our physical death enters its new mother's womb and when there, to see if the fetus is aware of its mother's feelings and emotions; and

-to investigate the subjects' encounters on other planets if and when such encounters occurred.

Horatio
O Day and night, but this is wondrous strange!

Hamlet
And therefore as a stranger give it welcome.
There are more things in heaven and earth, Horatio,
Than are dreamt of in your philosophy.

—William Shakespeare

13

He's a Scout. He Makes Maps

I basically followed the same procedure in my Second Research Project as with my First. I advertised once, and received seventy-three responses. Each respondent was sent a letter explaining the nature of my research, along with information concerning many misconceptions about hypnosis. They were also sent an Information Survey (Appendix A). Sixty people returned completed Surveys. I met with and held sessions with fifty of them.

One subject failed to go into hypnosis but of the forty-nine who did, forty-four regressed to what seemed to be past lives. In the regressions, I first took them to a lifetime of their own choosing. I did not suggest a time or place and did not generally spend much time there or expend much effort in probing that experience. I had found in my First Research Project that as the hypnotic regression proceeded, the subjects seemed to get more "into the experience" and concurrently deeper into hypnosis and accessing more detail.

Similar to my first research, the first past lives to which most of the forty-four subjects regressed, were generally quite mundane and ordinary, considering the periods and locations. The time periods ranged from 1563 to 1914—in those cases where time periods were obtainable. The locations when stated, were as varied as Holland, United States, Israel, Rome, France, England, Scotland and Australia. Examples of the situations in which the subjects found themselves: three were slaves, one a prisoner, one a spirit, one a Catholic bishop, one was the wife of shopkeeper, two were soldiers, and one was a monk. Most just lived and worked in rural settings.

However, two subjects who regressed to their first past lives in our sessions, encountered lives that I found most interesting. The first was Lucy, a fifty-year-old white, married lady with one child. She is a college graduate. She indicated on her Information Survey that she believed in reincarnation but that she did not believe in evolution. Her religious affiliation is Christian.

In the first past life to which she regressed, she found herself in a very dangerous situation. Lucy had regressed to being a twelve-year-old girl named Aleshia. I was trying to proceed in my usual manner to establish gender, race, clothes, family connections, etc., but Lucy was reliving the situation in which she found herself, and instead of responding directly to my questions, she was urgently trying to tell me what was going on.

RTJ: What's your name?

L: Aleshia.

RTJ: What's your last name, Aleshia?

L: I'm in an arena. I'm a Christian. (Obviously not responsive)

I continued right on in my usual manner, trying to establish names and other information.

RTJ: There's someone standing right next to you. This is a friendly person and they're talking to you. They're calling you by your name. Your first name and your last name. What do they call you?

Lucy would have none of it.

L: I'm a Christian. I'm in the arena. And there's a priest. And a cross. It's not a friendly place.

RTJ: Look around. What do you see? Why isn't this a friendly place?

L: I'm going to get killed.

RTJ: Why are you going to be killed?

L: I'm a Christian.

RTJ: Who is it that's going to kill you?

L: The Romans. There's a lion. And a priest.

RTJ: Is the priest going to be killed too?

L: He's looking at me.

RTJ: What's he saying to you?

L:	He's not saying anything.
RTJ:	Is the priest going to die too?
L:	I don't know.

I was able to establish that she had survived those games in the arena. Then,

RTJ:	How did you escape from that lion?
L:	Dead bodies.
RTJ:	Did the lion kill those people?
L:	Uh huh.
RTJ:	How did you escape without being killed?
L:	Dead bodies over me.
RTJ:	Were you under the dead bodies?
L:	Yes.

I was also able to establish that Aleshia lived to be in her twenties, with a mate, in Israel. She never had any children and died at the age of twenty-three. After her death in that lifetime, I took her forward in time.

RTJ:	Now I want you to go forward in time. You still haven't been born into another lifetime. Where are you now?
L:	I'm floating.
RTJ:	In this period in-between lives. Have you received any instruction?
L:	We're not allowed to say.

The reason I have included this last dialogue, is that this answer (We're not allowed to say) is not unique with only this subject. I have received similar answers from several subjects in response to my inquiry about instruction in their in-between lifetimes.

After taking Lucy back up on her safe cloud, I had her look back at all the people she knew in that lifetime. Did she know any of those people in her present lifetime? She responded that the priest in the arena is her mate in her present lifetime.

Lucy's experience in meeting again with an individual encountered in a past life is not unusual. This occurred numerous times with my subjects, and has been reported by others who use hypnotic regression techniques. See for example, the remarkable case uncovered by Dr. Brian Weiss reported in his 1996 book *Only Love is Real.*

The second subject who regressed to her first past life to a very interesting personality, was Frances. Frances, a sixty-three-year-old lady with grown children, is a college graduate with graduate degrees. She regressed while in hypnosis in her first past life to a time-period with information that could easily be verified. She easily went into a deep state of hypnosis. Her self-report hypnotic depth was eight.

Frances regressed to being Jessie Fremont, the daughter of U.S. Senator Thomas Hart Benton and the wife of John Charles Fremont. Jessie was born in 1824 and died in 1902. Her father, Senator Benton served as Senator from Missouri from 1820 to 1850. Jessie married John Fremont in 1841. Fremont rose to the rank of major general in the United States Army and died in 1890. He was and is still a very well known explorer of the West. I didn't know all of this detail, of course, before I regressed Frances.

I easily obtained the above information about the Fremonts and Senator Benton after the regression. Frances' regression illustrates the problems that are involved in trying to obtain information from a regression that will help determine whether or not a subject is actually regressing to a past life, or is somehow accessing information previously learned and now forgotten.

Frances was in a deep hypnotic trance and was well beyond the ability to knowingly fabricate a story for some agenda of her own. Whatever information she was accessing in her subconscious seemed real to her.

RTJ:	Tell me now, how old are you?
F:	I'm eighteen.
RTJ:	Look down at your arms. Tell me, what's the color of your skin?
F:	I'm white.
RTJ:	Are you a girl or are you a boy?
F:	A girl.
RTJ:	Look down at yourself and tell me what you are wearing.
F:	It's navy blue with white flowers on it...small white flowers. The skirt is rather full.

RTJ:	Look down at your feet. What are you wearing?
F:	Black shoes. They're rather high.
RTJ:	Look at the shoes. How are they fastened?
F:	They have laces…and…they have something…like you catch the string on them.
RTJ:	Tell me, what's your name? You're an eighteen-year-old girl.
F:	Jessie.
RTJ:	What's your last name?
F:	Benton.
RTJ:	Do you have a mate?
F:	Not yet.
RTJ:	Look around. Tell me. What do you see?
F:	I see books. It's my daddy's study.

Notice the expression "…my daddy's study". A rather juvenile expression for a mature, sophisticated sixty-three year-old lady.

In the book *Dream West*, author David Nevin recites that when Jessie was quite young, she formed the habit of rising in the morning before five o'clock. She made coffee, and took it to her father in his study. She spent many hours there while her father began talking out his ideas that he would later present in the Senate.

RTJ:	What's your daddy's name, Jessie?
F:	Senator Benton.
RTJ:	What town is this, Jessie?
F:	Washington.
RTJ:	What's your mother's name, Jessie?
F:	Iris?…Iria? Iris?

Here Frances raised her voice as if asking a question, rather than making a statement, indicating trouble in accessing whatever information might be in her subconscious. The reference material that I consulted disclosed that her mother's first name was Elizabeth. I was not able to ascertain whether or not Elizabeth might have had a nickname of Iris.

RTJ: What's your father's first name?

F: Thomas.

RTJ: What year is this, Jessie?

F: 18…1876?

Again note her answer was more like a question than a statement. I believe this date to be in error as she would have been fifty-two years old in 1876.

RTJ: Tell me, who is the President?

F: Lincoln?

Lincoln became President in 1861 and served until 1865. Jessie was alive during Lincoln's tenure, but Lincoln would not have been President in 1876.

RTJ: Now I want you to go forward in time, Jessie. Go forward 'till you're about 25 years old. Tell me, do you have a mate now, Jessie?

F: Yes.

RTJ: What's your mate's name?

F: Charles.

RTJ: What's Charles' last name?

F: Fremont.

In *Dream West*, the author confirms that Jessie and most others referred to John Charles Fremont as Charles.

RTJ: Do you and Charles have any children?

F: One.

RTJ: What's your child's name?

F: He's a little boy.

RTJ: What's your little boy's name?

F: Tom?

Jessie and Charles had six children in addition to one that was stillborn. The oldest boy was named John Benton Fremont, after both Jessie's husband and her father. He may or may not have been called "Tom" after her father.

RTJ: Where are you? You're twenty-five-.years old now.

F: St. Louis?

RTJ: What does your husband Charles do for a living?

F: He's a scout. He makes maps.

This is indeed accurate. Charles spent most of his adult life in the army exploring and mapping the West.

RTJ: Where do you live in St. Louis? In a house? Or an apartment?

F: In a hotel.

RTJ: What's the name of the hotel?

F: Jackson is all I see.

RTJ: What year is this now? You're twenty-five-years-old.

F: 1831? (Again, her response was more like a question.)

RTJ: Where were you born?

F: Virginia.

RTJ: Now I want you to go forward in time. This is some kind of a ritual or ceremony. Look around and tell me, what's going on?

F: It's an inauguration?

RTJ: Where are you?

F: In Washington.

RTJ: Who's being inaugurated?

F: Lincoln?

RTJ: Where are you at this inauguration? Where are you standing?

F:	I'm standing with my husband…and I'm talking with my father.
RTJ:	Where are you?
F:	We're near a doorway.
RTJ:	Is this outside or inside?
F:	It's inside.
RTJ:	What building is this in?
F:	I believe it's the White House. It's big.
RTJ:	Look around and tell me, what do you see?
F:	I see lots of people. I'm wearing a striped dress. It's pretty.

This could have very well been true. Jessie would have been thirty-seven years old when Lincoln was inaugurated, and she would have been married to Charles at that time.

RTJ:	Now Jessie, without any pain or fear, I want you to go forward in time to the day that you died in that lifetime. Look around. Tell me, where are you?
F:	It's a feather bed.
RTJ:	Is there anybody there with you?
F:	(Frances' voice is very soft now and she speaks hesitatingly.) I can feel a boy. Just like he's there.

One of her sons died at the age of four.

RTJ:	Now go right on through that death experience. You've just died in that lifetime. Tell me, can you see your body?
F:	Yes. I'm above it.
RTJ:	Is there anybody up there with you?
F:	(Tearfully.) Charles.
RTJ:	Has Charles already died?

F: Uh Huh.

Charles died in 1890. Jessie died in 1902.

I've presented the above portion of Frances' regression to illustrate some of the many problems inherent in the search for verifiable information for use in authenticating the regressed experiences. So many of the details of the existence and life of Jessie Fremont were confirmable by public records that the possibility of coincidence is just about eliminated. In a later interview, Frances indicated that she had heard of Senator Benton, and several years before she had read the popular book, *Dream West*.

We have a paradox here. If we humans have lived before? If Frances has lived as Jessie Fremont before her present lifetime, would it be unusual for her now to be drawn to books about the lives of Jessie, Charles, and Senator Benton? I suggest that it would not be unusual.

However, Frances's regression to the life of Jessie Fremont, although very suggestive of the phenomenon of a past life, is compromised in our efforts to establish verification *beyond doubt* because of the common and easily obtainable information of the existence and activities of Senator Benton, and Jessie and John Charles Fremont; together with Frances' prior contact with this information. The possibility of cryptomnesia can't be eliminated. This case is not nearly as conclusive as the cases reported by Tarazi and Brown, where the authenticating details were obscure and generally unavailable to the public.

In each of the subjects' first lives to which they regressed, Frances was the only one that regressed to a time period with information that could be verified. However later, when the subjects were regressed to the past lives they encountered just prior to their present lifetimes, things were different, as we will see.

I next took each subject back to the first time they had ever lived on Earth *in any form*.

Long before conducting the current research, while in College, I had studied the evidence for the evolution of our human species and I continue to do so. I am convinced of its validity. Even so, I was unaware of the adventure that lay ahead of me. I want to share that adventure with you in the next few chapters.

14

I Feed the Wolves

After the subjects had left their first past life, I put them on their traveling cloud and took them back to the first time they had ever lived on Earth *in any form.*

When so regressed, thirty subjects appeared to be Homo sapiens; that is, humans as we might know them today, even if living in very primitive cultures.

Two of the subjects who were "human" the first time they were on Earth, regressed to fairly modern times, and I would like to share their experiences with you first.

One young, single lady regressed to living as a young woman named Kara Duport, in Germany in the middle 1800s. Another young married lady with two children regressed to being a young man, living in rural Texas in the early 1800s.

Illustrative of those regressions to "human" existences the first time they were on Earth, are the following:

Louise is a forty-eight-year-old married lady with six children. She attended college for two years. On her Information Survey, she indicated that she believed in reincarnation, but did not believe in evolution. Louise was a good subject, easily going into a deep trance state. As we join Louise in her regression, I had just brought her down from her traveling cloud.

RTJ: Look down at yourself. What do you see?

L: It's cold. Cold. It's very cold.

RTJ: Look down at yourself. What do you see?

L: I have only furs.

Louise was very deep into hypnosis and was talking laboriously. I was tempted to lighten her state a little to help her speak more clearly, but decided to let her stay where she was momentarily so I could better access her subconscious.

RTJ: Are you human?

L:	Oh yes.
RTJ:	Are you a man or are you a woman?
L:	I'm a woman.
RTJ:	About how old are you?
L:	About ten. Ten.
RTJ:	Look down at your arms. What's the color of your skin?
L:	It's light
RTJ:	What are you wearing?
L:	It's fur.
RTJ:	What kind of an animal did the fur come from?
L:	It's a big animal.
RTJ:	Did you see the big animal killed?
L:	Oh…I helped.
RTJ:	How was the animal killed?
L:	I don't like it. It's so hard.

Louise was speaking and responding in a rather childish manner. In her personality as a ten-year-old child, raised in a very primitive culture, she was reliving the experience and responding with her own concerns, not necessarily with the information that I was seeking to elicit.

RTJ:	Who killed the animal?
L:	The men.
RTJ:	How did they kill the animal?
L:	They have rocks…and…spears. (Then excitedly.) It ran over. It fell.
RTJ:	Look at the animal. What does it look like? Does it have four legs?
L:	Oh yeah. It's a big animal.
RTJ:	Is it bigger than you?

L:	Oh yes. Now it's awful.
RTJ:	Look at its head. What does its head look like? Tell me.
L:	A long trunk or nose.
RTJ:	Do you eat the meat from the animal?
L:	Oh yes.
RTJ:	Now I want you to go forward in time. Go forward until you're about twenty years old in that lifetime. Tell me, do you have a mate now?
L:	I'm only fourteen.

Notice here, that although I "suggested" to her that she would go to the age of twenty, she didn't comply but responded in a different manner. This happens, often contrary to the criticisms of using hypnosis in past-life regressions. The subjects don't just follow the suggestions of the hypnotist.

RTJ:	Do you have a mate when you're fourteen?
L:	I died when I was fourteen.
RTJ:	How did you die? This is the day you died in that lifetime. How did you die?
L:	I fell. It's steep.
RTJ:	What's your name?
L:	Onga.
RTJ:	Unka? (I didn't quite understand her and mispronounced her name. She corrected me.)
L:	Onga (with emphasis.)
RTJ:	Now Onga, I want you to go right on through that death experience.
	Without any pain...without any fear. You've just died in that lifetime.
	Tell me, can you see your body?
L:	Uh Huh.

RTJ:	Now you've just died in that lifetime. Without any pain, without any fear.
	Where are you that you can see this body?
L:	I feed the wolves. (Then contentedly.) Oh I'm so warm now.

Apparently Onga's body was just left in the open without any burial or other disposition.

Another subject who regressed to a primitive culture, but still as a human, was Elizabeth. Elizabeth is a thirty-four-year-old married lady with two children. As had Louise, she also attended college for two years. On her Information Survey, she indicated a belief in reincarnation, but not a belief in evolution.

We join Elizabeth just after I brought her down from her traveling cloud. Elizabeth spoke hesitatingly, in a soft voice. Like many in hypnosis, she tended to just answer my inquiries fairly precisely, without volunteering much information. And again, as with Louise, once Elizabeth got into her regressed personality, she was reliving the experience and tended to tell me what was going on that was important to her, instead of directly answering my rather routine inquiries. This makes me have to dig and dig for information.

| RTJ: | Look down at yourself. What do you see? |

I ask the question this way to avoid suggesting any particular life form.

E:	I see a river.
RTJ:	Look around. Is there anybody by the river?
E:	Yeah. There's a few people there.
RTJ:	I want you to look at them real close now. What do they look like?
E:	They're not wearing any clothes. They just have…around their waist.
RTJ:	Are they wearing anything below their waists?
E:	Yeah. Like a fur.
RTJ:	Are you there by the river?
E:	Uh Huh.

RTJ: About how old are you now? How many seasons have you lived?

E: I don't know. (Long pause. Then.) I'm about fourteen.

RTJ: Are you a boy or are you a girl?

E: I'm a girl.

RTJ: Look down at yourself, are you wearing any clothes?

E: Yeah. I'm wearing a fur.

RTJ: Look down at your feet. Are you wearing anything on your feet?

E: No.

RTJ: While you're looking down. Look at your arms. Tell me, what color is your skin?

E: I'm brown.

RTJ: Look at those people beside the river. Are they carrying any kind of weapons?

E: They have a spear and a fish. (Then she volunteered.) They have lots of Fish.

RTJ: How did they catch the fish?

E: With the spear.

RTJ: Are those people any relation to you?

E: My family.

RTJ: Do you have any brothers and sisters?

E: Yes.

RTJ: Do you have a name? What do they call you?

E: Tora.

RTJ: Are your mother and father still alive?

E: No.

RTJ: What happened to them?

E:	They were killed.
RTJ:	What killed them?
E:	An animal.
RTJ:	What did the animal look like? Can you tell me?
E:	He had big teeth.
RTJ:	Was he bigger than they were?
E:	Yeah.
RTJ:	What did he look like? Did he have four legs?
E:	He looked like a tiger or a bear. Like a lion.
RTJ:	Now I want you to go forward in time. Go forward in time until you're about twenty years old. Now, you're twenty years old. Do you have a mate now Tora?
E:	Yeah.
RTJ:	Look at your mate. He's right there close to you. Does your mate have a name? (Long pause, but no answer). What do you call him?
E:	Elum. (Phonetic.)
RTJ:	Do you and Elum have any children?
E:	Four.
RTJ:	What's the name of your oldest child? (Long pause. No answer). Is your oldest child a girl or a boy?
E:	A boy.
RTJ:	He's standing right there close to you. What do you call him?
E:	I don't know. (Pause.) I have three girls too.
RTJ:	Now Tora, I want you to see yourself sitting down to a meal. Tell me, what kind of food are you eating?
E:	I'm eating meat.
RTJ:	What kind of animal did the meat come from?

E:	I don't know. It's a leg.
RTJ:	(I misunderstood her.) By a lake?
E:	No, it's a leg.
RTJ:	Did your mate kill this animal?
E:	Yeah.
RTJ:	Did you see the animal before it was skinned? Before he took the leg off?
E:	No.
RTJ:	Now Tora, I want you to go forward in time. Forward in time to some kind of a ritual or ceremony. Look around. Tell me, what's going on? What's the purpose of this ritual or ceremony?
E:	They're hungry. Nobody's got food now.
RTJ:	What's the purpose of the ceremony? What are you doing?
E:	(Very long pause). Somebody had a baby?
RTJ:	Look around Tora. What's the country look like?
E:	It's dark out. There's a fire?
RTJ:	When it's light. What does the terrain look like?
E:	There's rocks. It's green.
RTJ:	Are there any kind of dwellings? Can you see any dwellings?
E:	There is…in the sides of a hill.
RTJ:	Do you live in caves, or do you build in the side of the hill?
E:	We build in the sides of the hill.
RTJ:	Now I want you to look at your mate. He's standing right there close to you. Look at his hair. What color is his hair?

E:	Black.
RTJ:	Is it straight or is it pretty kinky?
E:	It's straight.
RTJ:	Look at his eyes. What color are his eyes?
E:	They're dark brown.
RTJ:	Are they round, or are they slanted?
E:	They're slanted.
RTJ:	Look at his face. Does he have any hair on his face?
E:	No hair.
RTJ:	Look down at his body. What is he wearing?
E:	It's a drape. On his hips.
RTJ:	Now without any pain or fear Tora, I want you to go forward to the day that you died in that lifetime. Look around, where are you?
E:	They're carrying me.
RTJ:	What happened to you?
E:	I fell off the cliff.
RTJ:	Now go right on through that death experience. You've just died in that lifetime. Can you see your body?
E:	I was getting sticks. And my kids were there. An animal came.
RTJ:	What kind of an animal is this? Look at the animal.
E:	Oh. Like a pig.
RTJ:	How big is it? Is it as big as you are?
E:	He's got tusks.
RTJ:	Does he have a lot of hair on him?
E:	He's got hair.
RTJ:	How big is he? Is he as big as you are?

E:	No.
RTJ:	How come you fell off the cliff?
E:	He came to hit me.
RTJ:	Now go right through that death experience. You've just fell off the cliff and died. Can you see your body?
E:	Yeah.
RTJ:	Where are you that you can see your body?
E:	I'm on the ground. They're covering me with things.
RTJ:	Are they burying your body?
E:	They going to burn me.
RTJ:	Where are you that you can see what's going on?
E:	I'm on the cloud.
RTJ:	Is this your traveling cloud?
E:	Yeah.

The final regression to a very primitive culture where the regressed personality was human contained a couple of experiences that to me, at least, were quite unexpected. Marie is a twenty-three-year-old married lady, with one child. On her Survey, she indicted a belief in reincarnation, and marked "uncertain" concerning a belief in evolution. Her self-report hypnotic depth was seven.

In her regression to the first time she lived on earth in any form, she regressed to being a sixteen-year-old girl, named "Entira". (Phonetic). She wore a simple tunic, with nothing on her feet. I had already established that she was a girl, and that her parents were still alive.

RTJ:	Now Entira, I want you to go forward in time. Forward in time 'till you're about twenty years old. Tell me, do you have a mate now?
M:	Yes, I do.
RTJ:	Look at your mate. He's standing right there close to you.
M:	She. (Entira corrects me.)

RTJ: She? Are you a man or are you a woman?

M: I'm a woman, but I have a female mate.

Entira then was cooking bread with her mate and later had a meal made of a corn mush. She stated that they grew the corn. She and her people lived in tent-like structures; skins stretched over wood.

RTJ: I want you to go outside the tent now. Go outside and look around. What do you see?

M: We're right by a large body of water.

RTJ: Do you see any animals?

M: Yes.

RTJ: What do you see?

M: I see a dolphin. She's my friend.

RTJ: Can you talk with this dolphin?

M: Mentally. We don't physically talk. This is impossible.

RTJ: Do you talk to your mate? Do you have a language?

M: Yes. But not something I can describe.

RTJ: Can you talk to her mentally?

M: Uh Huh.

Strangely enough, I had one other subject who, when she regressed to the first time she ever lived on earth in any form, talked mentally with a dolphin.

I want to next acquaint you with a few cases where the subjects regressed to life forms that were hominoid, but not quite human as we know it. I realize that having actually spoken with those sub-human personalities occurring in the past lives of my subjects was an experience that most of us have never had or could even imagine having. Come with me and have a look into our evolutionary past.

15

I Don't Live Anywhere

In compiling the data that I accumulated in my Second Research Project, I classified the life forms to which my subjects regressed as Human, Sub-human, Animal, Spirit, and Other. Five of my subjects regressed to what I have classified as sub-human. I wouldn't want to argue with an anthropologist about my classifications, as they are not intended to be scientific. I based my classification of sub-human on a number of different factors: the appearances of the personalities to which the subjects regressed, whether or not they had a language, had names, the primitive types of weapons used, their living conditions, their knowledge of the use of fire for heating and cooking, uncertain family relationships, and the types of animals that they hunted, killed, and ate, to name just a few.

One of my best subjects was Edith, a sixt y-year-old, retired, black lady. Edith was my first subject to regress to being a sub-human in my Second Research Project. Edith is married with one grown child. On her Information Survey she indicated uncertain as to a belief in some form of life after the death of the body, she was uncertain as to a belief in reincarnation, and she indicated a belief in evolution. Her self-report hypnotic depth was six.

Her regression was to a sub-human life form and to a culture that lacked even a simple dwelling in which to live, had little or no language skills, had very primitive weapons for hunting, and had no definite family relationships. Her culture did have the knowledge and use of fire.

She signaled that her traveling cloud had stopped and that she had reached the first time she had ever lived on Earth in any form.

RTJ:	Come down from your cloud now. Look around and tell me, where are you?
E:	It's kind of dark. There seems to be a lot of trees around.
RTJ:	Look down at yourself. What do you see?

E:	Hairy. Hairy looking. I think I'm a man.
RTJ:	Are you a human being?
E:	Ah. (Long pause.) Yes...but very hairy. Very hairy.
RTJ:	About how old are you? Can you tell me?
E:	I look like I'm thirty something, but I'm younger.
RTJ:	Do you have a name?
E:	No.
RTJ:	There's someone standing right next to you. Probably a friendly person. They're talking to you and calling you by name. What do they call you?
E:	(Long pause. No answer.)
RTJ:	Listen to them. You can hear what they're saying. Are they talking in a language?
E:	Just a couple of words. Just...pointing...and mostly a lot of hand movements.
RTJ:	Look at them. Describe them to me. Is this a man or a woman?
E:	A female.
RTJ:	What does she look like?
E:	Just like...hair...and it's very long.
RTJ:	Is she wearing any clothes?
E:	Just shabby things. It looks like leather. More like a suede thing. (Then) Hide.
RTJ:	Look down at yourself. Are you wearing anything?
E:	A fur looking thing. Some kind of animal fur.
RTJ:	Look at your arms while you're looking down..
E:	Hairy. Big. Hairy. Coarse hair. Lots of hair.
RTJ:	Look around now. What's the terrain look like?

| E: | Hilly…rocky and trees. Very tall trees. I can barely see the tops of them. |
| RTJ: | Can you see any kind of a dwelling? |

Long pause. No answer.

RTJ:	Do you live in a dwelling?
E:	I don't live anywhere.
RTJ:	I want you to see yourself. You're eating some kind of a meal or something. What are you eating?
E:	A bone. Chewing…getting meat off a bone.
RTJ:	What kind of an animal was this, that you're eating?
E:	I don't know.
RTJ:	Did you kill this animal that you're eating?
E:	(Here Edith had a puzzled look on her face. Then she answered) I have some kind of stick that I carry. (Which is not particularly responsive.)

In many of these cases that I have classified as Subhuman, for some of my questions, the regressed personalities didn't seem to comprehend what I was asking. Sometimes when this occurred, there would just be no answer, or their response was as though I had asked something different.

RTJ:	Describe this stick to me. Is it pointed?
E:	Yes.
RTJ:	Is there anything on the end of the stick?
E:	It's sharp.

Later.

RTJ:	Do you have a mate?
E:	I think that woman…(Then she volunteered.) She's very unpleasant.
RTJ:	Do you have any children with this woman?
E:	I see children around.

RTJ:	Are some of those children yours?
E:	I think so. It's hard to tell.
RTJ:	Look at one of the children. What does the child look like?
E:	It's a little girl. Her nose is kind of round. and flat like. Her face is flat like.
RTJ:	What color is her skin?
E:	Dirty. It's not too clean. The sun has been on it. She's not brown though.
	Sort of…we're not black people.
RTJ:	Is she hairy like you
E:	She has long hair, but not on her body. It's long, scraggly hair. Very unkempt. And bushy eyebrows. Everyone has very heavy eyebrows. And small eyes.
RTJ:	Do you have a language?
E:	Yes. Shouting and pointing. A lot of pointing.
RTJ:	Where do you get your food?
E:	We find it, sometimes. It just seems to be there.
RTJ:	Do you go hunting?
E:	Yes. They hunt at night.
RTJ:	What kind of animals do you hunt?
E:	They have tails. They look reptilian. I don't know. They're not big though.
	They're not too difficult to kill.
RTJ:	How do you kill them?
E:	Stick 'um.
RTJ:	Now I want you to go forward to some kind of a ritual or ceremony.
E:	There's a fire. They're cooking.

RTJ:	Now I want you to see one of these animals that you kill. How big is it? Is it as big as a child?
E:	Bigger. They're about as big as I am. But not as fat.
RTJ:	Does it have four legs?
E:	Uh huh.
RTJ:	What's its head look like?
E:	It has a long head, with a...it's a reptile. Yeah, it is. It has those eyes.

Those hooded eyes. You have to peel the skin off. |
RTJ:	Now I want you to go forward in time without any pain or fear. Go forward 'till the day you died in that lifetime. Look around now and tell me, where are you?
E:	Yes. It seems different. A tree fell on me and I can't get up.
RTJ:	Are there other people there with you?
E:	Yes.
RTJ:	How old are you now?
E:	I'm still about the same age. I didn't get much older.

Shortly after the regression session, I sent Edith a tape of the session (with induction omitted) as I did for all subjects who requested it. I also sent her photocopies of drawings of pre-historic animals from the library that seemed to be similar to the description of the reptiles they hunted. Edith responded immediately indicating that the Komodo dragon was the reptile they hunted.

The Komodo dragon, although of very ancient origin, still exists in Indonesia. A large, lizard looking reptile, it can grow to ten feet in length and weigh up to 300 pounds. It generally feeds on monkeys, pigs, and deer. [30]

My second subject who regressed to a sub-human culture was Dianna, a twenty-seve-year-old white married lady with one child. On her Survey, she reported a belief in reincarnation but that she did not believe in evolution. Her self-report hypnotic depth was six.

She regressed to being a female, wearing skins and barefooted. As I was trying to establish an identity, the following occurred:

RTJ:	Tell me, what's your name?
D:	I don't have one.
RTJ:	How old are you? How many seasons have you lived? Do you know?
D:	No.
RTJ:	Are you a younger person or an older person?
D:	I look old.
RTJ:	Now I want you to see yourself sitting down at a meal. What are you eating?
D:	Berries.
RTJ:	Are you eating anything else?
D:	Raw meat.
RTJ:	What kind of animal does the meat come from? You can see the animal now.
D:	Big.
RTJ:	What does it look like?
D:	Like an elephant.
RTJ:	How did you get the meat from the animal? Who killed the animal?
D:	They did. The men.
RTJ:	How did they kill it?
D:	With sticks.

I wasn't able to establish with this subject if the elephant was like a modern elephant, or more like a mammoth or possibly the predecessor to the mammoth. And later.

RTJ:	Do you have a mate?
D:	Lots of them.
RTJ:	Do you have any children.

D:	Yes. Boys.
RTJ:	How many boys do you have?
D:	Four.
RTJ:	Were they from different mates?
D:	I don't know.
RTJ:	Do the boys have names? Did you give them names?
D:	No.
RTJ:	Tell me. How do you communicate with the boys?
D:	Grab them.
RTJ:	Do you have a language that you speak to them?
D:	No.
RTJ:	Look around. Do you have any dwellings there where you live?
D:	In the dirt.
RTJ:	Describe one of these dwellings to me. What do they look like?
D:	They're in the side of the hill.
RTJ:	Is this kind of a cave?
D:	Yeah.

Dianna's regressed life-form died by falling while she was being chased by a big animal.

My next subject who regressed to being a sub-human with a concomitant culture, was Lucy. You will remember Lucy from Chapter Thirteen. Lucy is a fifty-year-old white woman, married with one child. She also reported on her Survey that she believed in reincarnation but not evolution.

She regressed to being a young female, dressed in skins, with brown-colored skin.

RTJ:	What's your name?
L:	Don't have a name.

She reported that she had a mate.

RTJ:	Is this your mate you're looking at?
L:	Uh huh.
RTJ:	Does he have a name?
L:	No.
RTJ:	How do you communicate with your mate? Do you have a language?
L:	He shows me.
RTJ:	Now look at your mate. Does he have any weapons?
L:	Yes.
RTJ:	What kind of a weapon does he have?
L:	A stone.
RTJ:	Now I want you to see yourself eating a meal. What are you eating?
L:	Raw meat.
RTJ:	What kind of an animal did the meat come from?
L:	A deer.
RTJ:	Who killed the deer?
L:	He did.
RTJ:	How did he kill the deer that you're eating?
L:	With a stone.
RTJ:	Are there any dwellings that you see?
L:	No.
RTJ:	Where do you live?
L:	Inside of mountains. It's dark.
RTJ:	What kind of weapons are they carrying?
L:	Stones. (Then.) There's a lance. With a pointed end.

RTJ:	What's the pointed end made out of? Is it made out of anything except just wood?
L:	It's not wood.
RTJ:	What is it?
L:	It's stone.
RTJ:	How big is this lance?
L:	It's bigger than them. They're not very big.
RTJ:	What kind of animals are you hunting?
L:	Elephants.
RTJ:	Look at the elephants. Do they have a lot of hair on them?
L:	Yeah. Big ears. They have hair on their ears.

This elephant would seem to be more mammoth than a modern elephant.

Lucy had three children, but they had no names. She died "'cause I'm old" in the cave where she and her mate lived.

My fourth subject who regressed to a sub-human personality was Jeanne, a thirty-four-year-old woman, college graduate, married, no children, who believes in reincarnation but is "uncertain" about evolution. She regressed to being a brown-skinned woman, wearing no clothes, who had lived about thirty seasons. She had a mate:

RTJ:	Does he have much of a forehead?
J:	No. Not too much.
RTJ:	Is he wearing any clothes?
J:	Uh huh.
RTJ:	What's he wearing?
J:	Like a skirt.
RTJ:	What's the skirt made out of?
J:	Some kind of leopard.
RTJ:	Is he carrying any kind of a weapon?

J:	Just a stick.
RTJ:	Do you see any kind of dwellings?
J:	Just mostly caves.

And then,

RTJ:	Now without any pain or fear. I want you to go forward in time to the day that you died in that lifetime. Look around. Where are you?
J:	At the bottom of a mountain.
RTJ:	Why are you dying?
J:	I fell down. On the rocks.
RTJ:	How old are you now? How many seasons have you lived?
J:	About forty.

My last subject who regressed to being sub-human the first time she ever lived on earth in any form, was Frances. You will recall Frances from Chapter thirteen, she is a sixty-three-year-old lady, with graduate degrees. On her Survey, she reported a belief in both reincarnation and evolution.

RTJ:	Come down from your cloud now. You're coming down slowly and gently. Look down at yourself. What do you see?
F:	I'm bipedal.
RTJ:	Look down at yourself. Are you in human form?
F:	Well, sort of.
RTJ:	Look down at your arms.
F:	Lots of hair on 'um. Light brown.
RTJ:	Are you male or female?
F:	I'm female.
RTJ:	About how old are you? How many seasons have you lived?

F: twenty?

Frances answered by raising her voice like a question, indicating uncertainty.

RTJ: Look around. Do you see other people?

F: Yeah.

RTJ: What do you see?

F: They look like I do.

RTJ: What do they look like? Look at the one right next to you.

F: It's got a strange head.

RTJ: Does it have hair on its head?

F: Uh huh.

RTJ: Look at its body. Does it have any clothes on?

F: No. Lots of hair.

RTJ: Is this a male or female?

F: It's male. Muscular.

Frances was such a good hypnotic subject, I thought I would see if she would speak in another language.

RTJ: Now your mate is standing very close to you. You're talking to him. You can hear yourself talking to him. What are you saying to him?

F: Let's go over in this direction.

RTJ: Now listen very carefully now. You can hear yourself saying that to your mate. "Let's go over in this direction." I want you to repeat to me now, what you're saying to your mate. In the same language. Can you say that?

F: It's mostly grunts. (It didn't work.)

RTJ: Now I want you to see yourself. You're eating a meal or something. What are you eating?

F: Berries.

And they also were eating;

RTJ: Do you ever eat meat?

F: Occasionally.

RTJ: What kind of animal does the meat come from?

F: Other people.

She and her mate lived in a cave and had a fire at the entrance. I tried to get her to describe her people.

RTJ: Now I want you to see yourself. You're in the cave. There's probably a fire at the entrance to the cave. Your mate is probably right there by you.

                              Look at his head. Does he have much of a forehead?

F: No.

RTJ: What's the shape of his head? Describe it to me.

F: (Long pause.) The hair comes straight off into the face. There's not much along the sides.

RTJ: Look at the back of the head. Is the back pretty good sized?

F: Uh huh.

A member of Frances' family who is a professional artist, at Frances' direction, made a drawing of Frances' mate in that lifetime. I've attached a copy of this drawing at the end of this Chapter. This drawing is much more illustrative of the life form to which Frances regressed than does the above dialogue.

Next, I want to explore with you the rest of those regressions to the first time the subjects ever lived on earth in any form. I still occasionally replay the tapes of the following regressions with awe, trying somehow to fit these regressed life-forms into a revised, personal concept of the nature of our universe. It may be, as Shakespeare wrote in Hamlet: "There are more things in heaven and earth, Horatio, than are dreamt of in your philosophy." Bear with me with an open mind.

16

I'm Not From Here

You may have had problems with some of the preceding regressions, particularly with those regressions where the regressed personality was something less than a human as we would presently define and recognize a human. But as I previously indicated, I will present my subjects' regressed experiences, without any attempt on my part to suppress or withhold those that might seem outrageous, or too "far out" to some readers.

However, before presenting the rest of the regressions where the subjects regressed to life forms that I was not able to classify as either human or subhuman, I would like to point out what I believe to be a valid principle of logic in evaluating these experiences.

The overall purpose of my two research projects was to investigate the phenomena of normal, healthy adults, regressing, while in hypnosis, to what seem to be previous lives. I, like many, have been raised and educated in the traditional Western culture. Because of that, there may be a temptation to accept some of the regressions that appear "normal" within the framework of our heritage, but to reject others as being too "far out."

For example, one of my subjects regressed to a past life as a woman, born and raised on a farm in middle America, who never married, stayed with her father after the death of her mother, and died of old age on that same farm. Although I was raised in Middle America, I can't point specifically to anyone who I personally knew who fit that scenario, but we all know that that this life-style has happened over and over again, many times. It fits the reality with which we are familiar.

Even Carol's regressed experience as a Viking on the East coast of what is now England (Chapter ten), isn't too "far out", as we all know that the marauding Vikings did settle there. Her experiences as Yon could very well have happened without stretching our credulity.

Some, or possibly all of the following life forms experienced by my subjects may not fit into the reality which many of us generally know, and some readers may find rejecting those experiences necessary, even after accepting some of the more mundane regression experiences that do fit into our traditional belief-systems.

If, in using hypnosis, I am actually accessing memories in a subject's subconscious which I am confident that I am, then what I am accessing may be true memories, or possibly something else. But to accept some experiences because they fit into life styles with which we are generally familiar in our culture, and reject others because they do not, seems illogical. I believe such acceptance and rejection, based on the nature of the experienced regression, is not helpful or logical in evaluating the phenomena of past-life regressions.

But then, how do we evaluate the cases that may seem "far out" to many, even though we may feel comfortable with those that fit well within our pre-conceived beliefs as to the nature of reality?

We can reject the entire notion of our having past lives, as do many of our fundamental religious friends. But then what do we do with the "hard-core" cases of Stevenson, Tarazi, Rick Brown and others (Chapter twelve)? Can it be that we are now getting glimpses of a different realty with which we are not familiar? A real look into our evolutionary past? With this in mind, look at the more difficult cases.

Two of my subjects regressed to what are unquestionably animal life forms. The first was April, a thirty-seven-year-old black, married lady with two children. On her Information Survey, she indicated "uncertain" as to her beliefs in (1) some form of life after the death of the body, (2) reincarnation, and (3) evolution. She was a very good hypnotic subject, easily going into a deep state of hypnosis; eight on the self-report depth scale.

After putting April on her traveling cloud, she first regressed to being a black, female slave named Leatha. After briefly exploring that lifetime, I then took her back to the first time she ever lived on Earth in any form. The following transpired:

RTJ:	Look down at yourself. What do you see?
A:	I'm a cat. Big cat.
RTJ:	What kind of a cat is this?
A:	It's a lion. No hair.
RTJ:	Are you a male or a female lion.

A: Male.

RTJ: Look around. What do you see? What's the terrain look like?

A: Trees.

RTJ: Can you see other lions?

A: Over there. (Gesturing). In the trees. Lot of big trees.

I start out to explore April's life as a lion, but,

RTJ: Now I want you to see yourself, you and whatever lions are with you you're eating a meal and,

She interrupts,

A: Getting ready to fight. He's looking at me.

April doesn't even attempt to respond to my question. She is reliving her life as a lion and she is reporting what is important to her at that moment in her life.

RTJ: Is this another lion?

A: The same lion.

RTJ: Who's he going to fight?

A: Me.

RTJ: Are you afraid of him?

April shakes her head no.

RTJ: Do you have a fight with this other lion?

A: He's walkin'. He's still looking but he's walkin' down that way.(Pointing.)

RTJ: All right. He's going away. Now I want you to see yourself going hunting.
                   You're hunting for food. Tell me, are there other lions with you?

A: Don't see any more.

RTJ: What are you hunting?

A: To the trees. I see a lot of birds.

Not particularly responsive.

RTJ:	What else do you see?
A:	A lot of vines…and grass…and there's a pathway right through here.
	(Gesturing). And there's a stream…over there.
RTJ:	Do you see any other kinds of animals?
A:	Monkeys over there. (Pointing.)
RTJ:	What kinds of animals do you usually hunt?
A:	Deer.
RTJ:	Have you ever seen any humans around?

April shakes her head no.

RTJ:	Now I want you to go forward in time in that lifetime. Without any pain or fear. This is the day you died in that lifetime. Look around. Where are you? What do you see?
A:	Darkness.
RTJ:	Why are you dying in that lifetime?
A:	I don't know. Something that I can't see.
RTJ:	Now without any pain or fear, go right on through that death experience.
	You've just died in that lifetime. Can you see your body?
A:	I'm laying down in a field.
RTJ:	Where are you that you can see your body?
A:	I'm above it.
RTJ:	Is there any body up there with you?
A:	I'm just standing above it.

The sub-human regressed personalities that I encountered in Chapter fifteen fit nicely into my understanding of our human evolutionary past, particularly from recent pre-human remains discovered in Africa. However, April's regression, and the regression experienced by Elaine, take us back much further in time.

Elaine is a white, forty-nine-year-old divorced mother of two. She is a college graduate with graduate degrees. She too was a super hypnotic-subject. In hypnosis, she was in a very deep state with a self-report depth of nine.

She first regressed to being a fifteen-year-old girl, living in 1891 in Kansas City. Upon regressing to the first time she ever lived on earth in any form:

RTJ:	Look down at your self. What do you see?
E:	Leathery. (Speaking hesitatingly and very softly.)
RTJ:	Are you a human being?
E:	No.
RTJ:	What are you, can you tell me?
E:	I'm a lizard or something.

Elaine described herself as a juvenile lizard of uncertain gender. She described herself and the other lizards around her as having four legs, long tails, large jaws, with only the birds as their enemies. They reproduced by the females laying eggs. They ate grass and bugs, lived on a rocky terrain near water, and she eventually died from a lack of water.

After her death, I took Elaine forward in time to the lifetime she lived just before her present lifetime. She regressed to being a twenty-five-year-old girl, with brown skin, living in a mud hut in a desert country.

In the post-hypnotic interview, Elaine told me that her mother lizard was very large; larger than a human. It was surprising that she didn't seem amazed that one of her regressions was as a reptile. Once having the experience, she just accepted it. At my request, Elaine drew me a sketch of what she looked like as a reptile. I have attached a copy of her sketch at the end of this Chapter.

She also told me that the questions I was asking seemed different than what was important to her as she was going through her experiences. As an example, she said that while in the last lifetime when I was asking her how many children and grandchildren she had, that she could see and feel them, but to stop and count them seemed unimportant. They were just her family. Several subjects have told me that when I was asking for information to develop their identities and to explore their past lives, they felt my questions were intrusions into the experience.

Elaine also told me that when I put her back up on her traveling cloud and took her to another lifetime, that she immediately went there, and was waiting for me to catch up. Several other subjects have told me the same thing. Elaine was the second person to tell me, that as I took her from lifetime to lifetime, she could see her other lifetimes but we went right past them. She said, "We just seemed to flip through them."

My next two subjects who regressed to something other than human were a forty-five-year-old single lady named Kay, and a fifty-nine-year-old retired school teacher named Vicki. On her Information Survey, Kay indicated that she believed in life after death, believed in reincarnation, but was uncertain about evolution. Vicki also answered her Survey by stating that she believed in life after death, believed in reincarnation, but did not believe in the evolution of humans from other species.

Both Kay and Vicki went deep into hypnosis and regressed to being bodiless spirits. When I asked Kay to come down from her cloud, the following occurred:

RTJ: Come on down from your cloud now. Look down at yourself. What do you see?

K: I'm changing form.

RTJ: Describe the forms to me. What are you changing to and from? (Long silence.) What does it look like?

K: Rocks and plants.

RTJ: Are you part of the rocks and plants?

K: When I want to be.

RTJ: Look down at yourself. Do you have a body?

K: When I'm a rock.

RTJ: Do you have a choice when you're changing from rocks to plants? Why do you change from rocks to plants?

K: Because it feels good to be a rock or a plant, or in-between.

RTJ: Look around and tell me, wherever you are. What do you see?

K: Rocks and plants, and, you call them sprites, I think.

Later I ask her,

RTJ: Where do these forms come from? Do you know?

K: They just are.

RTJ: Now go forward in time to the time that you leave that place. How do you leave that place? Do you die?

K: I just left the planet.

RTJ: Is that planet Earth?

K: Long ago.

When I brought Vicki down from her traveling cloud, and ask her to look down at herself and tell me what she saw, she replied:

V: Flashes of light.

RTJ: Do you seem to have a body?

V: No.

RTJ: Can you sense others there with you? Are there people there?

V: No. Well, everything's there, but it's not people.

RTJ: What are they? Look at them?

V: Flashes.

RTJ: Do you feel the presence of any kind of intelligence?

V: Oh, surrounded by it. It's all comfortable.

RTJ: Is this intelligence communicating with you in any way?

V: It just is.

RTJ: Do you feel that you have a greater understanding of the nature of reality from this intelligence?

V: Yeah.

My last subject that had a different past-life form was Alice, a thirty-three-year-old married lady, one child, with two years of college. She is partly Korean in ancestry, born, raised, and educated in the United States. On her Survey, Alice indicated a belief in life after death, reincarnation, and evolution.

Alice's regression is one of the most interesting that I have ever encountered. Whatever we may conclude took place in her regression, it was very emotional and was very real to her. Her words and concomitant emotions had the characteristics of a real event. I realize that for those who have not been present during a hypnotic regression such as that experienced by Alice, to appreciate the emotional intensity of what the subject is experiencing is difficult.

After bringing her down from her traveling cloud:

RTJ: Look down at yourself. What do you see?

A: It's dark. It's different. (Then very emotional.) I'm not me.

RTJ: What do you see when you look down at yourself?

A: A short, short person. (Then she exclaimed.) Not a person.

RTJ: You're not a person. Look at yourself and tell me, what are you?

A: I'm not from here.

RTJ: Where are you from?

A: Some where else.

RTJ: Look down and tell me what do you see? Do you have arms and legs?

A: Yes.

RTJ: Are you in human form?

A: Kind of.

RTJ: Look down at your arms. What color is your skin? Do you have skin?

A: It's gray. Long arms.

RTJ: Are you wearing any kind of clothes?

A: No.

RTJ: Are you male or female?

A: I'm not either.

RTJ: Is there anyone else around that looks like you?

A: No. (Then with emotion.) They left me.

RTJ: Why did they leave you?

A: I don't know.

RTJ: What did they look like before they left you? Do they look like you?

A: Yes.

RTJ: You can still see one in your mind. What do they look like? Are they tall or short?

A: Short like me.

RTJ: What color is their skin?

A: Gray, like mine.

RTJ: Do they wear any clothes?

A: They have on helmets. I don't.

RTJ: How did they come and go? How did they leave?

A: (She starts...) They left in a gray...(Then, sobbing.) They pushed me down the ladder.

RTJ: Was this some kind of a flying machine?

A: Yes. (Alice is still sobbing.)

RTJ: Now you've just been pushed down the ladder. Look up at the machine.
        What does it look like? What shape is it?

A: It's long. And oval kind of. (Crying) They're leaving.

RTJ: Why are they leaving here? Do you know?

A: (Very emotional.) I don't know.

RTJ: Where are they going?

A: I don't know.

RTJ:	Where did you come from?
A:	My home. (With strong emphasis.)
RTJ:	Is your home on this planet.
A:	No. (With anguish).
RTJ:	What's your name? What do they call you?
A:	Sh...sh. (She tries to say something. Then...) I don't know.
RTJ:	Now tell me. If you're not male or female. How do your people reproduce?
A:	They make us.
RTJ:	Do you have a mother and a father?
A:	No.
RTJ:	How do they make you?
A:	There's a big box. They put something in it.
RTJ:	You can watch the ship as it's going up. But you're not feeling any pain now. Watch this ship as it takes off. Tell me, does it have engines?
A:	You can't see anything.
RTJ:	What makes the ship go up?
A:	Centrifugal force.
RTJ:	Does it go up to another planet?
A:	Not now.
RTJ:	Where is it going? Do you know?
A:	No.
RTJ:	Now I want you to go forward in time in that lifetime. Go forward just a few years now. Tell me, where are you now?
A:	In the forest.

RTJ:	What are you doing there?
A:	I'm just here. I'm shriveling.
RTJ:	Are you getting old?
A:	I'm dying.
RTJ:	Why are you dying?
A:	I'm not suppose to be here.
RTJ:	Now without any pain or fear, go right on through that death experience.
	Now you've just died in that lifetime. Can you see your body?
A:	Yes.
RTJ:	Now look down at yourself. You've just died in this lifetime. You're not feeling any pain now. What's your body look like?
A:	It's just like a little seed.
RTJ:	What happened to your body?
A:	It shriveled up. It dried up.
RTJ:	Where are you that you can see this body?
A:	I'm above it.
RTJ:	Is there anyone there with you?
A:	No.
RTJ:	Now I want you to go forward in time now. I want you to leave that scene.
	You still haven't been born into another lifetime yet. Look around. Where are you now?
A:	I'm on the cloud.
RTJ:	Is this your traveling cloud?
A:	Yes.

My research had not only discovered humans with past-life memories as animals and spirits, but now Alice with a past life as an alien from another planet. I initially thought that these findings were unique, but I soon found that I was mistaken.

Psychiatrist John Mack in his recent book *Abductions. Human Encounters with Aliens,* reported a couple of his patients experienced past lives in which they also had an alien identity.[31] Others, usually therapists using past-life therapy with clients, have also reported encountering alien past-lives.[32]

In October of 1994, I sat as an observer in a therapy session in Los Angeles with a therapist working with a middle-aged lady who recalled in a hypnotic regression that she had been abducted by aliens several times during her present lifetime. She also has regressed to a past life where she had an alien identity.

Certainly those regressions to animal and more primitive life forms are consistent with accepted, scientific concepts of evolution. That all organisms now found on Earth evolved and descended from a single origin of life through natural selection, is accepted by most scientific disciplines even though unaccepted by some religious fundamentalists.[33] Even knowing this and intellectually accepting it, I found the actual face-to-face contact with those alien and primitive life forms startling.

Although natural selection of life forms is random, by chance; these life forms evolved and developed in an environment of established laws of physics and biology. My question isn't how did humans come into existence, but how or by what means did these environmental laws come into being? Also unanswered in the evolutionary concept, are the existence of human consciousness and the source of that portion of our individuality that survives our physical death.

The foregoing and following regressions should help us all ponder our place in nature and in the universe.

To further compound the matter, in my next Chapter, I will report to you on the regressions of several of my subjects who describe past lives on other planets.

17

Like a Fish with Wings

In my First Research Project, after taking the subjects through two past-life experiences of their own choosing, and then through the in-between life periods, I put them back on their traveling cloud and returned them back to the present time and place. On the cloud coming back, just as a curiosity on my part, I told the subjects that it was night time, and they could look up in the sky and see all the stars and planets. I then ask if they had ever lived on another planet. Most surprisingly to me, thirty of my eighty-one subjects who had regressed to what seemed to be previous lives responded that they had lived on another planet.

In the post-hypnotic interviews, a number of my subjects who initially recalled living on another planet, commented that they wondered what it was like on that other planet. I didn't pursue those experiences in that first research, as I already had a large number of variables to examine. However, in my Second Research Project, I did pursue the other-planet experiences, with startling results.

Of the fifty subjects in my Second Research Project, forty-four regressed beyond birth to what seemed to be lives they had lived before. Of those, thirteen responded that they had lived on a planet other than Earth. Of those thirteen, ten were able to regress to their lives on those other planets.

If you have followed the advances made in astronomy as reported from time-to-time in the popular magazines and journals, it's apparent that a majority of the world's scientists do believe in the probability of extraterrestrial life existing somewhere in the incredible size of our universe. Concurrent with such belief, is the realization that such life forms, if they exist, evolved in different environments and possibly, even probably, in different forms from our own, and that some of those life forms are probably both younger and older and are more or less advanced than our own culture.

If their lives on other planets as experienced by my ten subjects are accurate, then a variety of life forms exist in our universe other than humans as we know them on Earth. Illustrative of one such life form is that experienced by Jenny, a forty-two-year-old white, married lady with no children. She has had three years of college. On her Survey, she indicated a belief in some form of life after death and reincarnation, but not in evolution. She was in a medium-deep state of hypnosis as we progressed through the regression:

RTJ:	Now the cloud is starting up again. It's bringing you back to the present time. It's bringing you back to Colorado Springs. It's Thursday, late in the morning. But it's night time up on the cloud. You can look up in the sky and see all the stars and the planets. Tell me, Jenny, have you ever lived on one of those planets?
J:	Uh huh.
RTJ:	When did you live on one of those other planets?
J:	Long time ago.
RTJ:	Were there people on the planet with you?
J:	Uh huh.
RTJ:	What did they look like? Describe one to me.
J:	They're short. Very short. Two feet. (Then she volunteered.) The planet's wonderfully blue. Two moons.
RTJ:	Look at one of the persons and describe them to me. What's the color of their skin?
J:	Greenish gray.
RTJ:	Do they have hair?
J:	No.
RTJ:	Do they have eyes?
J:	They have visual apertures. I wouldn't call them eyes.
RTJ:	Do they have a mouth?
J:	More like a beak, but, yeah, a mouth. It's flexible.
RTJ:	Do they speak to each other?
J:	Yes. (Then again she volunteers.) Soft hands. Very long fingers. Very soft.

RTJ:	Do they reproduce like humans do on Earth?
J:	No.
RTJ:	How do they reproduce?
J:	More like plants.

My experience with Jenny caused me to realize that in regressing subjects back to their lives on other planets I would probably encounter very unique life forms. But Maxine surprised both of us.

Maxine is a fifty-nine-year-old retired, white, single lady, with one adult child. She indicated on her Survey that she believed in life after death and reincarnation, but was uncertain about evolution. Her life form on the other planet surprised me, but was unbelievable to her.

RTJ:	Tell me, have you ever lived on one of those other planets?
M:	Uh huh.
RTJ:	Just for a moment now, I want you to go back to that other planet where you lived. Look down at yourself now. What do you look like?

Here Maxine starts laughing, almost hysterically. When I got her quieted down, she answered:

M:	I'm like a fish. But not a fish, you know. I'm like a fish with wings.
RTJ:	With wings?
M:	Uh huh.
RTJ:	Now look down at yourself. Do you have arms and legs?
M:	Kind of. They're very short and stumpy.
RTJ:	What color are you? Do you have a skin?
M:	Pink.
RTJ:	Look around. Are there other people there with you?
M:	Yes.

RTJ:	Do they look just like you?
M:	Yes.
RTJ:	Look at them. Do they have heads?
M:	Uh huh. I've got a head.
RTJ:	Eyes, a nose and a mouth?
M:	Well, fish like.
RTJ:	Are you under water?
M:	No, it's a very, very heavy gas.
RTJ:	Tell me, how do you reproduce? How do you bring new people into the world?
M:	With eggs.
RTJ:	Now I want you to go right back up on your cloud.

Here, Maxine settles comfortably back in her chair, smiles contentedly, and says:

| M: | It was a nice place. |

Strange as it may seem to us in our present lives, apparently the feelings and emotions Maxine had from that unusual life were happy.

After the session, at my request, Maxine drew a sketch of what she looked like on that planet. I've attached a copy of her sketch at the end of this Chapter.

Two of my subjects when they regressed to lives on other planets found themselves in spirit form, without physical bodies. Interestingly, with one who was in spirit form, I had the following dialogue:

RTJ:	Have you ever lived on one of those other planets? (Long pause.) You can tell me.
T:	Not suppose to.
RTJ:	You're not supposed to tell me?
T:	They can't know yet. It's for the future.

I tried to pry,

| RTJ: | Why aren't you suppose to tell me? |

T:	It's not ready yet.
RTJ:	Can you look back and see what you looked like when you lived on one of those other planets? Can you tell me what you looked like? (Long pause.)
	Do you have the same kind of body that you now have?
T:	Oh no, no, no.
RTJ:	What did it look like?
T:	I didn't have a body.
RTJ:	How do you know that you lived on one of those other planets?
T:	I'm not suppose to say.

I found this dialogue interesting because I had encountered similar responses ("I'm not suppose to tell you") while probing the in-between-lifetimes of a couple of my subjects in my first research.

I continued on, trying to work with one subject each day. My next subject was Martha.

RTJ:	Now I want you to see yourself when you lived on this other planet. Look down at yourself. What do you look like?
M:	I have very long limbs. Like hinged. Segmented. Kind of hinged. I kind of shuffle forward and backward, or sideways. (Pause.) Our sun. Kind of orange.
RTJ:	What's the terrain look like? Do you see any trees?
M:	It's hard, crusted sand. Smooth rocks. Hot all the time.
Later..	
RTJ:	Now there's someone standing right next to you. What color is their skin?
M:	Everything's orange.

In the post-hypnotic interview, Martha said, "That other planet place. That was really strange." When I ask her how they reproduced themselves, she answered, "Segments break off."

Another subject who regressed to another planet was Janice.

RTJ:	I want you to see yourself on this other planet. Look down at yourself.
	What do you see? What do you look like?
J:	(Stating incredulously.) I think I'm green.
RTJ:	Are you male or female?
J:	It doesn't matter.
RTJ:	There's some of these people that are like you standing right next to you.
	Look at them. Tell me, what do they look like? Are they green too?
J:	Uh huh.
RTJ:	Look at their bodies. What do they look like?
J:	They're like…skinny.
RTJ:	Do they have legs and arms?
J:	Uh Huh.
RTJ:	Do they have a head?
J:	Uh Huh.
RTJ:	Do they have eyes?
J:	Kind of.
RTJ:	Do they have a mouth and a nose?
J:	Unh unh. No nose.
RTJ:	Now I want you to see yourself. You're eating something on that planet.
	What do you eat?
J:	Bugs.

In the post-hypnotic interview, I asked Janice as I did with all my subjects, to give me her immediate reaction as to whether or not what she experienced was real, or did she think she just imagined the experiences. She replied, "I saw them.

I don't know whether or not they were real. I question whether I had really been on another planet. But it really seemed real."

When I commented that the life forms she saw were strange looking, she said, "Oh they were. They didn't scare me though." I asked, "How tall were they? Were they as tall as you are now?" She replied, "No, I think they were shorter, and (she started laughing) they had four arms. Two on each side. And they were real thin, but kind of big heads. But they didn't necessarily seem out of proportion. But I think I'd been there more than once. It was like I had been there before I was here."

When Alice regressed to a life on another planet, she had a number instead of a name.

RTJ:	Do you have a name?
A:	(Hesitatingly.) Yes.
RTJ:	What's your name?
A:	It's a number.
RTJ:	What's your number?
A:	Four four three six.
RTJ:	Why do you have numbers instead of names?
A:	To keep track of us.
RTJ:	Look down at yourself. Are you wearing any clothes?
A:	Yes.
RTJ:	What are you wearing?
A:	A one-piece suit.
RTJ:	What color is this suit?
A:	Kind of a muted green.
RTJ:	Look at some of the other people around there. What do they look like? Do they have hair.
A:	Don't have hair.
RTJ:	Do they have eyes and a nose and a mouth?

A:	Yes.
RTJ:	Do they have teeth?
A:	Yes.
RTJ:	What's the terrain look like?
A:	It's cement. Clean.
RTJ:	Do you have any children?
A:	No.
RTJ:	Do you have a mate?
A:	No.
RTJ:	How do your people reproduce? How do they produce children?
A:	Like humans do.

My subject Jean didn't have a number instead of a name, but she was hairy.

RTJ:	Now just for a moment, I want you to go back to that planet. Look down at yourself. What do you see? What do you look like?
J:	Stringy. Long, long slender hands. Long fingers.
RTJ:	Are you a male or a female?
J:	Female.
RTJ:	Do you have arms?
J:	Yes.
RTJ:	What color is your skin?
J:	Greenish brown. It's hairy.
RTJ:	Are you wearing any kind of clothes?
J:	No.
RTJ:	Look around. Are there other people like you there?
J:	Uh huh.

RTJ:	What do they look like?
J:	Hairy.
RTJ:	Do you see both men and women?
J:	You can't tell the difference.

Two of my subjects, Martha and Janice, reported that they have lived on other planets more than once. While they were experiencing their past lives on other planets, I asked them, "When did you live on one of those other planets?" They both reported, "Different times."

An obvious problem arises here that may tend to stretch your credulity. Not only do life forms on other planets appear to exist that don't look like us, but also, those portions that survive those existences, are a part of our human past.

In addition the problem arises that these surviving personalities, whether we call them spirits or souls, or whatever, somehow are transported or transport themselves from planet to planet. How does this happen? We know that the distances from Earth to other stars and galaxies is measures in light-years; the time it would take light, traveling at about 186,300 miles a second, to reach them. The technologies with which we are familiar indicate for physical life as we know it to travel those distances and survive is highly unlikely.

But can we assure ourselves that the present known laws of physics do in fact correctly describe reality, at least for that intangible part of ourselves that survives our physical death? And therefore, that the experiences described in these regressions were fantasies, delusions, some type of distortions of reality, or that they simply didn't take place?

Is it possible that our cultural concepts of reality are too limited? Or do these experiences give us insights into alternative realities and forms of intelligent life of which we have not even dreamed?

Because of the deep trance these subjects were in, I was convinced that I was interacting with their subconscious minds, by-passing their conscious minds, and that the experiences I encountered were just not made-up stories.

For most of my subjects, these experiences caused them to ponder the nature of the universe and their place in it. It did for me also.

18

The Smell of the Gas Killed me

Nearing the close of my personal inquiry into the past-life phenomena, I would like to report on several miscellaneous matters of interest I encountered in my Second Research Project.

Three of my subjects regressed in their lifetimes just prior to their present lifetimes, to being individuals who were put to death for being Jews.

One forty-five-year-old woman, who is Jewish in this lifetime, regressed to being a twenty-five-year-old Jewish woman named Nadia Shuman, living in Berlin in the years 1936-7. She died of starvation in a German concentration camp.

Another fifty-year-old woman, a Christian in her present lifetime, regressed to being a twenty-year-old Jewish woman named Jeron Christofsky (phonetic), living in Germany in 1940. Being Jewish, she wore a bright colored star on her clothing. She died by being gassed in the "showers" in a German concentration camp. After her death in that lifetime, while still in hypnosis, she commented, "The smell of the gas killed me." Not many people around today know how the gas smelled that the Germans used in the "showers" to kill Jews.

A third woman, not Jewish in her present lifetime, regressed to being a six-year-old Jewish boy named Isaac Brokowitz (phonetic), living in Poland in 1930. He was shot to death along with a number of other Jews. He stated that the men who shot them were in uniform and had stars on their hats.

One last experiment that I conducted with most of my subjects in my Second Research Project, was while the subjects were still in hypnosis after their deaths in their past lives just prior to their present ones, I took their surviving spirit into the womb of the woman who was to be their new mother. I was attempting to explore when the surviving spirit entered its new mother's womb. And when there, was the fetus aware of its development and was it aware of its mother's feelings and emotions.

The spirit of one subject entered its new mother's womb at the time of conception. For ten subjects, the fetus had not assumed human form when the spirit entered its new mother's womb. The spirits of twenty-five of my subjects entered the fetus when the fetus was in human form in the womb, and one spirit joined the fetus at birth.

Of those who entered the womb in some manner, thirteen were aware of the fetus' development as it occurred, and thirty were aware of their mother's feelings and emotions during pregnancy. Interestingly, one spirit entered a fetus and then withdrew as the fetus aborted, but then re-entered a new fetus in the same mother at a later time.

These results are certainly consistent with and support the controversial concepts that the mind of the fetus is aware, that it records memories, and that it is affected by the emotions and feelings of its mother.

One other regression that was particularly unusual, was that of a forty-one-year-old man, married, college graduate, who regressed to being a pilot, killed in a crash along with others in his craft. Superficially, this may not seem unusual, but follow the dialogue as it develops:

RTJ:	Come on down from your cloud now. Tell me, how old are you?
Subject:	I don't know.
RTJ:	Are you a young person or an old person?
Subject:	Old.
RTJ:	About how many seasons have you lived? How many years?
Subject:	I don't know.
RTJ:	Look down at yourself. Look down at your arms. What color is your skin?
Subject:	Whitish.
RTJ:	Are you wearing any clothes?
Subject:	It's like a flight suit.
RTJ:	Are you wearing anything on your feet?
Subject:	Yes. Boots. Silver boots.
RTJ:	What's your name?

Subject:	Sh…sh…(Subject struggles to speak, but seems unable to pronounce the word.)

I follow my usual technique to try to get a name.

RTJ:	There's someone standing right next to you. This is a friendly person.
	They're talking to you. Listen. You can hear them talking. They're calling you by your name. What are they calling you?

But the subject is excitedly following his own agenda.

Subject:	Everyone's yelling. Some kind of ship and it's out of control.
RTJ:	What kind of a ship is this? Is it on the water?
Subject:	No.
RTJ:	An airplane?
Subject:	Some kind of a ship and it's out of control.
RTJ:	What are they calling you? What's your name?
Subject:	Shappi. (phonetic.)
RTJ:	Why are you in this plane, Shappi?

Again, the subject isn't responding to my question, but is trying to tell me what is going on.

Subject:	I think we have an engine problem.
RTJ:	Are you a pilot in this plane or a passenger?
Subject:	Pilot.
RTJ:	What kind of a plane is this? How many engines does it have? Look around. You can tell me.
Subject:	I don't see any engines.
RTJ:	How is the plane propelled, Shappi?
Subject:	There's a panel in front of me. I'm trying to control it but it's loosing.

RTJ:	Look down out of the plane. What do you see down below?
Subject:	Ground.
RTJ:	Where are you? What's the terrain look like?
Subject:	New Mexico.
RTJ:	What are you doing in this plane? Why are you in this plane?
Subject:	We were watching something.
RTJ:	Does the plane have windows? Can you look out?
Subject:	Yeah.
RTJ:	Look out. What's the shape of the plane? Does it have wings?
Subject:	I can't tell. I'm looking out the front.
RTJ:	Are there other people in the plane with you?
Subject:	Yes.
RTJ:	Look close at the person who is next to you. What do they look like? (long pause.) Do they have human form?
Subject:	Oddly.
RTJ:	Do they have a head?
Subject:	It's…it's oval.
RTJ:	Do they have eyes, Shappi?
Subject:	Big eyes.
RTJ:	Are they men or are they women?
Subject:	Men.
RTJ:	Are you a man?
Subject:	Yes.
RTJ:	Where's your home, Shappi? Where are you from?

Subject:	It's crazy.
RTJ:	Where are you from?
Subject:	Artoos (phonetic.)
RTJ:	What kind of energy propels your plane?
Subject:	Sort of electromagnetic.
RTJ:	How long have you been flying in your plane?
Subject:	Long time.
RTJ:	All right, now I want you to go forward in time a little bit. What happens to the plane? Do you crash?
Subject:	It crashes. I keep seeing…there's a control board.
RTJ:	What's the control board look like? Describe it to me.
Subject:	It's curved. There's a plate the size of a hand. Switches. Buttons.
RTJ:	Do you know why the plane was out of control?
Subject:	I keep saying we can't balance it out.
RTJ:	Are you hurt in the crash?
Subject:	We're all killed.
RTJ:	Now go right on through that death experience. You've just died in that lifetime.

The subject interrupts.

Subject:	We're trying to eject. There's a ship inside the ship and we're ejecting.
RTJ:	But you didn't make it?
Subject:	We're too close to the ground. Some of us got out, but we're hurt.
RTJ:	Are you still alive?
Subject:	Barely.

RTJ:	You're in New Mexico. Do you know what town you're close to?
Subject:	There's no towns close. There's sheep.
RTJ:	Do you die there?
Subject:	Yes.
RTJ:	Now you've just died in that lifetime. Can you see your body?
Subject:	I can't breathe.
RTJ:	Go right on through that death experience. Can you see your body?
Subject:	I feel myself leaving.
RTJ:	Where are you going? Do you know?
Subject:	Just up.
RTJ:	All right, now before you get away. I want you to look down now. You can see the whole plane now. What's the shape of the plane?
Subject:	Oval.
RTJ:	Were the other people in the plane all killed?
Subject:	Yes.
RTJ:	Can you see their bodies?
Subject:	Some. Some are still inside. There are people there. I see people coming out in a jeep. I see two or three people in the jeep.
RTJ:	I want you to stay there looking down. What do the people in the jeep do? You can watch them.
Subject:	Two don't want to get out. The third...just walks towards the craft.
RTJ:	How are they dressed? The people in the jeep.
Subject:	Khaki pants.

RTJ:	What do they do now? You can watch.
Subject:	He's looking at the bodies, and he's puzzled.
RTJ:	Go forward in time now just a little bit. What do they do? What do they do with the bodies?
Subject:	I see military trucks coming.

The subject seemed quite disturbed at this point so I took him back up on his traveling cloud. The experience resembles somewhat the Roswell, New Mexico incident that some reports say occurred in July of 1947.

On the traveling cloud coming forward to the present time, as with all my subjects, I had him look up in the sky, look at all the stars and planets, and tell me if he had ever lived on another planet. He replied, "I'll say yes, but it sounds crazy."

In this subject's Information Survey, he reported that he was not suffering from any illness, was not in therapy, not taking any prescription medicine, and in all respects he appeared to be normal. In the post-hypnotic interview, he had difficulty accepting that his experience in the flying saucer was real. He did report that he had sometime in the past, read about the reported Roswell, New Mexico flying-saucer incident

Of interest, he also told me that as a young adolescent, he believed that he didn't belong here and that late at night, he would go outside with a flashlight and look up, wondering if someone was going to come and get him. From the published literature, such childhood experiences are not unusual for persons who regress while in hypnosis, to a past life with an identity as an alien. When I asked him if he wanted a copy of the tape of his regression, he replied, "I'm not sure I would want my wife to hear the tape. She'll think I'm crazy."

Again, as with Alice (Chapter fifteen), we not only have a human who regressed to what seems to be a past life, but was also an alien from another planet.

The fact that we appear to be spiritual beings having different life-form experiences gives rise to some very profound questions that I wish to address in the next two chapters

19

Our Surviving Entity

Before concluding this personal journey, I would like to add some very brief comments and speculation on the nature and origin of that portion of our individuality that survives our physical deaths.

In each instance, not only in what I refer to as the "hard core" cases of others, but also in each of my very many cases where the subjects have regressed to past lives, some portion of those past life-forms survived physical death and were reborn again. The evidence contained in these hard-core cases and in my own cases certainly indicates that a portion of our individuality survives our physical deaths, has lived in the past in other life forms, and will probably live again. We are indeed spiritual beings having lived many times before, and are now having another human experience.

In view of these facts, the question arises as to the nature and origin of that part of our individuality that survives our physical deaths. It is an object, a substance, a separate, individual, conscious, entity, even though not composed of physical particles and not occupying physical space.

This substance does not belong to "us." It is "us." Although separated from our physical bodies at our physical death, it nevertheless carries with it the totality of our past experiences. All as shown by the observations and experiments described in this book.

It would seem to be some form of energy, but probably not a form of the electromagnetic energy loose in our Universe that scientists regularly deal with, or it would have been observed and measured before now.

Traditionally, many theologians, philosophers and others, refer to this surviving entity as the soul, or our spirit, or our vital essence, or like terms. In this Chapter, I will refer to this surviving entity as a *subtle energy body*, this avoiding the traditional terms, all of which carry many preconceived beliefs that I wish to avoid. *Subtle*, meaning difficult to perceive or understand, and an *energy body* meaning some individualized power source functioning as the identifiable, surviving, spiritual unit that is "you", "me", and everyone else.

Religious philosophies such as Hinduism, Buddhism, Gnosticism, Theosophy and others that refer to the evolution of the subtle energy body, generally do so in a karmic sense: that is, that we exist in a moral universe; that there exists a form of cosmic accountability and justice, allowing us to learn and profit from our many lifetimes as humans, and possible to evolve to "higher" beings, or to suffer retribution. They characteristically do not deal with the subtle energy body before our species became human in our present form.

I take no issue with this karmic concept and it does give reasons for our continued rebirth in human form. However, I have seen no empirical evidence for it. It seems to be mainly based on faith. In saying this, I am not in any way trying to ridicule or denigrate those who do accept spiritual matters on faith. I like to think that I have an open mind on all spiritual doctrines, but accepting faith as a criterion for the search for truth I do not find helpful.

Current Christian thought would generally postulate that the subtle energy body was newly created by some divine source at the time of physical birth or conception. Other modern religions seem to have similar beliefs or at least, only deal with the evolution of the subtle energy body after it vests in the human body.

Many early philosophers gave thought to the evolution of the subtle energy body, but almost always in the sense of its progress through repeated lives as humans. The Greek theologian Origen (A.D. c.185-c.254), as did many others of his time, taught the preexistence of the soul. Origen added to preexistence the concept that the human soul "...is immaterial, and therefore had neither beginning of days, nor the end of life...."[34] He apparently did not consider the possible evolution of the human soul from lesser-life forms, but of course, this was long before Darwin.

The only direct reference before the modern era to the possible evolution of the subtle energy body in Darwinian terms with which I am familiar, is that of a Unitarian Minister James Freeman Clark, who in the 1800s stated:

"It is true that the Darwinian theory takes no notice of the evolution of the Soul, but only of the body. But it appears to me that a combination of the two views would remove many difficulties which still attach to the theory of natural selection and the survival of the fittest."[35]

An indirect reference in the late 1800s to the evolutionary similarity of both physical bodies and subtle energy bodies was made by American Poet Walt Whitman, who said;

> "…there is a distinction between one's mere personality and the deeper Self (and that) by means of metempsychosis and karma we are all involved in a process of spiritual evolution that might be compared to natural evolution." [36]

In modern times in 1996, Dr. Thelma Freedman discusses the possible connection between evolution and the human soul and states "…there are excellent grounds to believe that evolution really happens, and we can place reincarnation snugly within it." [37]

It is almost universally accepted in most scientific disciplines, that we humans in our present physical form (Homo sapiens), evolved from lesser life-forms. The fact of human physical evolution is sustained by a large body of empirical evidence.

Upon accepting that we are spiritual beings having a human experience, just the knowledge that our physical bodies have evolved over the ages from lesser-life forms seems quite incomplete. As a matter of simple logic, I would agree with Clark and Freedman and submit further, that to arrive at an understanding of the nature and origins of our subtle energy bodies, we *must* also begin with their evolution commencing at the time of the physical evolutionary origins of the human species.

The ability to explore these spiritual origins is exceedingly difficult, but possibly not impossible. Hypnotic regression experiments, although not producing readily verifiable evidence of the evolutionary nature and origins of the subtle energy body, never-the-less can give us possible insights, or at least some evidence of the existence of subtle energy bodies at very early levels of life.

As examples, in the hypnotic regressions that I conducted as a part of my second research project, many of which have been set forth in the preceding pages, each subject was asked to return to the first time they have ever lived on Earth "in any form." The subjects were not informed before the sessions that this "first time on Earth in any form" would be investigated; only that pre-birth regressions would be attempted.

As I mentioned in Chapter fourteen, five subjects regressed to being hominoid in form, but not human. Two were in spirit form, two were clearly animals. and one was an alien from another planet. Thirteen subjects did regress while in hypnosis to lives that they had lived on another planet at some time in the past, although not necessarily the first time on Earth.

I took all of my subjects who regressed to past lives to their physical deaths in those lifetimes regardless of the life-form to which they had regressed. All of those subjects, including those who had regressed to animal and alien lives, survived their physical deaths and rose-up out of their physical bodies, and they were able consciously to look down at their physical remains. Thus, indicating the survival of their subtle energy body.

The concept that is obvious here, and which may seem strange and be the most difficult for some to accept, is that our evolutionary past as spirits, animals, and even aliens could be real.

In discussions with well educated persons in scientific pursuits, I find that many of them are firm believers that in our vast universe there are indeed other forms of life, evolving in different environmental conditions than those on Earth. But when confronted with the regression of Maxine (Chapter seventeen) who lived as a fish-appearing life form in an atmosphere of heavy gas, or Alice (Chapter sixteen) who had one past life-form as an alien from another planet, many found these experiences too "far out" to seriously consider.

So then, if you the reader, find the evidence that we are spiritual beings in a human experience very compelling as I now do, how do we evaluate these strange experiences?

It is important to note that only a very few of my subjects knew each other both in my 1994 research, and also in my earlier research, where I worked individually with over 100 subjects. In each case, in both research projects, the subjects who regressed back beyond birth rose-up above their physical bodies after their physical deaths, and were able to consciously look down at their physical remains. They were all later reborn in human form. The similarity of these experiences among strangers at least eliminated the possibility that these incidents derive from some form of communication between the subjects.

Obviously, these regressions to life-forms other than human cannot be verified, and certainly standing alone, do not "prove" that our subtle energy bodies evolve along with our physical bodies.

The fact that there is a "hard core" of cases where hypnotically remembered human past lives have been well verified, does lend credence to the legitimacy of the entire past-life regression phenomena, even those to lives other than human.

Whatever the truth is concerning the nature and origin of our human subtle energy bodies, we now at least know it is far more involved, than (1) the traditional view still prominent in many Western religions that each subtle energy body is created new at the time of each human birth, or (2) some "scientific" concept that the matter and energy that comprise our physical bodies existing within the framework of the space-time continuum constitutes the only reality of human existence.

20

For What Purpose?

I suggest that it is not the major premise of past-life regression research that *all* past-life memories are true. Can possibly in a hypnotic past-life regression, subconscious material other than true past-life memories surface as real events? I think so. The important point in determining whether or not we, or most of us, or just some of us, have lived before is; are *any* of the memories of past lives obtained in a hypnotic trance, or recovered without the use of hypnosis, true memories of past lives?

It would seem fairly conclusive in reviewing the verified reports of Stevenson, Tarazi, Brown, and others, (the "hard-core" cases), that *some* subjects did indeed live before their present lifetimes.

In the prior chapters I have dwelt mostly on cases where hypnosis was used as a convenient and practical technique to facilitate regression, but the past-life phenomena is not inextricably linked with hypnosis. Many cases exist where hypnosis was not used with the same results. The well verified case of an English lady Jenny Cockell, detailed in her 1993 book *Across Time and Death*, and recently shown on a television documentary, is an excellent example of a verified past life where hypnosis was not a factor.[38]

Some dogmatic traditionalists hang onto the concept that the matter and energy that comprises our physical bodies within the framework of the space-time continuum, constitutes the only reality of human existence. They seem to conform to the belief that their community sets limits beyond which if you are "scientific", you cross at your peril. They should be encouraged. They have made great progress in their limited field to the benefit of all of us.

If we were examining research into subatomic particles, such as quarks instead of past lives, such would be labeled by such traditionalists as "real science"; an inquiry into the basic structure of the universe. However, if in fact we, or at least some of us, have had past lives as at least the hard-core cases pretty conclusively show; isn't this as much a part of the structure of the universe and as much a part of our reality as subatomic particles? Is its investigation not "real science" as well? I think so.

A common concept used in scientific research is the principle advocated by fourteenth century philosopher William of Ockham (Ockham's Razor) which is; that among competing theories, the simplest theory that embraces and explains the most known facts, is probably the true one.

The existence of past lives; reincarnation, is the best and most logical interpretation of these hard-core cases. There is no significant body of thought that anything else is happening.

Although the belief in past lives is many centuries old (hardly "new age"), such belief has traditionally been based, like the major religions of today, upon religious scriptures, divine revelation, and faith, without verifiable, empirical, data. Today's scientific methods realistically require that the proponents of a theory or proposition such as the existence of past lives, assume the responsibility of demonstrating the truth of their assertions. That burden has now been met.

The recall of regressed past-life memories is no longer a matter of religious belief or dependent upon the insights of the enlightened, but at least with *some* subjects, it is a matter of demonstrable, human experience. With the well-investigated and reported cases mentioned above, the burden of proof has shifted to the detractors of past lives to show that such reports are not true.

In Chapter twelve I attempted to discuss different criticisms of the past-life phenomenon, especially involving the use of hypnosis to facilitate regressions to what seem to be previous lives. In that Chapter I was able to eliminate (in my mind at least), any fraud or deception, genetic memory, ESP, or spirit possession as possible causes of the phenomenon.

When a subject is in a good medium or deep state of hypnosis, a generally increased responsiveness to suggestion occurs. In my research, after inducing hypnosis, I placed the subjects on a traveling cloud and did suggest to them that they were traveling back in time, before they were born. Is it possible that merely suggesting to the hypnotized subjects that they are going back before birth is enough to stimulate their subconscious minds to fantasize a prior lifetime? If so, how does this suggestion result in the subject recalling true information not known to either the subject of the hypnotist?

As my statistics show, not all of my subjects responded to these suggestions while in hypnosis and did not so regress. Those that did regress, seemed not only to recall past lives, with its attendant detail (identity, family, location, etc.) but many appeared to *relive* their experiences, expressing emotions of fear, love, even terror, and the emotions seem consistent with what they were experiencing

In the past lives encountered by my subjects reported in this book, no mental illness or other apparent psychological or emotional facts can account for the regressed experiences. Whatever your conclusions or inclinations may be concerning the past-life phenomenon, my research has shown that the regression experience is a common occurrence among normal, healthy adults. It is not a unique or unusual experience. The past-life phenomena are really not about belief; they are about experiences.

The reversal of the race and sex that the subjects encountered would seem to tend to bind us all together. No matter what our present race, religion, or economic condition may be, you and I most likely were, in one of our many past lives, persons of a different sex, race, and economic condition. And you and I may be different again the next time around. We really *are* connected to all of humanity.

The reversed roles encountered in past lives certainly give emphasis to psychologist Carl Jung's *anima*, the feminine side of the male, and the *animus*, the masculine side of the female. Jung maintained that deep inside of every male is a soft, feminine counterpart, and deep inside every woman, a masculine self. The regression experiences show how these archetypes could possibly come about; not through inheritance from ancestors, but from our own past-life experiences.

My regressions along with the past-life regressions in the hard-core cases, give us important insights into the process of death. My subjects in both projects experienced a total of 294 past lives, and importantly, 294 past deaths. Are we physical bodies with a spirit, or are we really spiritual beings with a succession of physical bodies? In all of the 294 past deaths experienced by my subjects along with those subjects in the hard-core cases, some vehicle that maintained a specific identity has survived their deaths; something real yet non-material. Call it a spirit, soul, or whatever appeals to you. I prefer to call it a subtle energy body as explained in Chapter nineteen.

Shedding the physical body through death distills that subtle energy body into its purest form, with an existence of its own in a nonphysical realm or dimension. I am now convinced that we indeed are spiritual beings, now having a human experience.

As mentioned in Chapter thirteen, my subject Lucy and others repeatedly met the same individuals in past lives. Other researchers have had the same experience. With some six billion people now inhabiting the Earth, with probably all or most having had past lives, the probability of meeting our present friends and mates in another life by chance is too small to seriously consider.

This gives rise to some very profound questions. Is there meaning in our inter-personal relationships? What or who choreographs these meetings? And to what end? Do we really have Soul Mates? What is the purpose of all this?

Do psychological and physical symptoms follow some of us from life to life as indicated by Dr. Brian Weiss's book *Through Time and Healing?* With no genetic connection from one life to another, how could this happen? How about sexual identity confusion or child prodigies?

The past-life experiences where subjects regressed to the first time they ever lived on earth in any form, and the animal and alien experiences as reported in preceding Chapters, all tend to show that we really are not separate from the rest of the universe, but are an integral part of it.

Is the exploration of the past-life phenomena complete? Of course not. It's only beginning. A new frontier. It will have to be improved upon from time-to-time as new discoveries are made in understanding what it means to be fully human. It's a work in progress. And a fascinating work it is. When a large num-ber of those in the so-called scientific community agree that the part-life phe-nomena is a proper subject of investigation, more rapid progress can and will be made.

I encourage you to be an open-minded participant, not just a by-stander in this adventure.

21

Epilogue

Why did I write this book? As I have reiterated in several of the foregoing chapters, my research was a personal journey to see if there really was any validity to the past-life phenomena. My inquiry was not to prove anything to others, and was not for the purpose of publication. Then why the book? And why the numerous published articles that I have authored? Is all this just some form of self-aggrandizement, or some sort of ego fulfillment?

I approached this inquiry as an agnostic and as a skeptic; not as a believer or necessarily a disbeliever in a divine presence. I was brought up in a traditional Middle American home, with traditional Protestant religious beliefs. I eventually rejected these religious beliefs, not because of any special traumatic events, but because I simply did not see that the verifiable evidence sustained those beliefs. I also rejected the concept that I should accept these traditional beliefs on "faith" with no sustainable evidence.

To my surprise, I discovered that the very large body of hard-core, verifiable evidence, not visions, sacred books, divine revelations, or the opinions of mystics, is that a portion of our individuality does indeed survive our physical death, has lived before, and will probably live again. We are, individually and as a species, spiritual beings, having lived many times before, and are now having another human experience. It has taken me a lifetime to discover these facts and arrive at these conclusions and to be aware of a spiritual existence for me and for all of us. Not really a new concept, but new for me.

Historians and philosophers have often told us that in order to really discover who we are, and what our place in the universe is, we must study the past. I would submit, that individually and as a species, in order to really grasp and understand who we are, we must also be aware of and explore lives that we have lived before, including the origin of that portion of our individuality that survives our physical death.

In writing this book and my numerous articles I am not on a mission, trying to save souls or save and reform society. But I believe that it is important, at least for me, that this information be available not only for anyone else who may be interested, but hopefully to be available in some form for me in my next time around.

This might seem very selfish and it possibly is. But, it has taken many years for me personally to reach these conclusions and I would like to think that in my next life time I would have this information available and be able to start at a higher level of awareness than I did this time.

So now I retire, *sine die,* until the next time.

Footnotes

Introduction

1. Stevenson, Ian. (1974). *Twenty Cases Suggestive of Reincarnation*. Charlottesville: University Press.
_____(1974). *Xenoglossy*. Charlottesville: University Press.
_____(1977). Research into the Evidence of Man's Survival After Death. *Journal of Nervous and Mental Disease*. 165 (3), 152-170.
_____(1980). *Cases of the Reincarnation Type*. Charlottesville: University Press.

2. Wambach, H. (1978). *Reliving Past Lives*. New York: Harper. See also
_____(1979). *Life Before Life*. New York: Bantam.

3. Tarazi, L. (1990), An Unusual Case of Hypnotic Regression with some Unexplained Contents. *The Journal of the American Society for Psychical Research*. 84, 409-425.

4. Brown, R. (1991). The Reincarnation of James, The Submarine Man. *The Journal of Regression Therapy*. VOl., 1, 62-71. This particular case was chronicled on the TV program "Unsolved Mysteries."

Chapter One.

5. See Ernest R. Hilgard's *Psychology in America, a Historical Survey*, for a more complete history of hypnosis. (1987). San Diego:Harcourt Brace Jovanovich, Publishers.

6. LeCron, L. M. (1968). *Experimental Hypnosis*. New York:Citadel Press.

7. Clark, R. L. (1995). *Past Life Therapy. The State of the Art*. Austin, TX:Rising Star Press.

Chapter Two.

8. Elman, Dave. (1964). *Hypnotherapy*. Glendale, CA:Westwood Publishers.

9. Basically in this research I used the Tart Self-Reporting Depth System for checking on depth. See Tart, C. T. (1970). Self-Report Scales of Hypnotic Depth. *The International Journal of Clinical and Experimental Hypnosis*, Vol. XVIII, 2, 105-125.

Chapter Three.

10. In Appendix B, I have set forth many of the more interesting statistics concerning the composition of my 104 subjects.

11. I graded the responses to my susceptibility test as (1) No response, (2) Light response, (3) Medium response, (4) Excellent response, and (5) Negative response.

Chapter Five.

12. Moody, R. A., Jr. (1975*)*. *Life After Life*. Covington, GA: Mockingbird Books.

13. Moody, Raymond A. Jr. (2000). *The Last Laugh*. Charlottesville,VA:Hampton Roads.

14. Stevenson, Ian; Cook, Emily Williams; & McClean-Rice Nicholas. (1989-90). *Omega*, vol. 20(11) 45-54.

15. Morse, M. & Perry, P. (1990*)*. *Closer to the Light*. New York: Villard Books. See also Osis, K. & Haraldsson, E. (1977). *At the Hour of Death*. New York: Avon Books.

16. Fiore, E. (1978). *You Have Been Here Before*. New York: Ballantine Books.

Chapter Six.

17. For a good up-to-date discussion of the nature of hypnosis, see Yapko, H. D. (1990). *Trancework*. New York: Brunner/Mazel.Also of value is Bowers, K. S. (1976). *Hypnosis for the Seriously Curious*. New York: Norton.

Chapter Seven.

18. Gallup, George, Jr. (1982). *Adventures in Immortality*. New York: McGraw-Hill.

Chapter Eleven.

19. Cornwall, I. W. (1968). *Prehistoric Animals and their Hunters.* New York: Praeger.

Chapter Twelve.

20. Stevenson, Ian.(1984). *Unlearned Language. New Cases in Xenoglossy.* Charlottesville: University Press. Many other past-life remembrances of children, some well verified, are set out in his 1997 book, *Where Reincarnation and Biology Intersect.* Westport, Conn:Praeger.

21. Stevenson, Ian. Xenoglossy: A Review and Report of a Case. *Proceedings of the American Society for Psychical Research.* (1974). Feb. Vol. 31.

22. Stevenson, Ian. (1976). A Preliminary Report of a New Case of Responsive Xenoglossy: The Case of Gretchen. *The Journal of the American Society for Psychical Research.* 70 (1), 65-77.

23. Wambach, H. (1978). *Reliving Past Lives.* New York: Harper.
_____(1979*). Life Before Life.* New York: Bantam.

24. Tarazi, L. (1990). An Unusual Case of Hypnotic Regression with some Unexplained Contents. *The Journal of the American Society for Psychical Research,* 84, 409-425.

25. Brown, R. (1991). The Reincarnation of James, The Submarine Man. *The Journal of Regression Therapy,* 1, 62-71.

26. Fiore, E. (1978). *You Have Been Here Before.* New York: Ballantine Books.

27. Weiss, Brian. (1992). *Through Time Into Healing.* New York: Simon & Schuster. See also Woolger, Roger J. (1987*). Other Lives, Other Selves.* New York: Bantam.

28. Stevenson, Ian. (1997). The Explanatory Value of the Idea of Reincarnation. *The Journal of Nervous and Mental Disease,* 165(5), 305-326.

29. James, Robert T. (1993). Regressed Past Lives and Survival after Physical Death: Unique Experiences? *The Journal of Regression Therapy.* VII, (1), 33-50. A

report on my second research project was also published in the same Journal. Vol. IX, (1), December, 1995.

Chapter Fifteen.

30. Steele, Philip. (1991). *Extinct Reptiles.* New York: Watts.

Chapter Sixteen.

31. Mack, John E. (1994). *Abductions. Human Encounters with Aliens.* New York: Charles Scribner's Sons.

32. Steiger, Brad. (1996). *Returning from the Light.* New York:Signet.

33. Mayr, Ernst. (1991). *One Long Argument.* Cambridge:Harvard University Press.

Chapter Nineteen.

34. Head, Joseph & Cranston. S.L. (1997). *Reincarnation:The Phoenix Fire Mystery.* New York:Julian Press.

35. Head, supra.

36. Head, Supra.

37. Freedman, Thelma. (1996). Two Notions:Snugging Into Paradigms. *The Journal of Regression Therapy.* X, 1, 38–42.

Chapter Twenty

38. Cockell, Jenny. (1993). *Across Time and Death.* New York: Fireside.

APPENDIX A

Information Survey

The following completed Information Survey was obtained from all subjects before their hypnotic sessions. Their answers were coded into a raw data-file along with information obtained from the hypnotic sessions for analysis.

Robert T. James
1983 North Academy
Colorado Springs, Colorado 80909
(719) 597-8803

Information Survey-Confidential

General

Please complete the following and mail in the enclosed envelope.

1. Name:_____
       (First)   (Middle)   (Last)

2. Address:_____
       (Street)   (Apt)

       (City)   (State)   (Zip)

3. Telephone:_____

4. Occupation:_____   5. Age____years

6. Sex: (1) Male_____ (2) Female_____

Education

7. High School:

       Graduated: (1) Yes_____ (2) No_____

       Number of years attended:_____

8. College:

       Graduated: (1) Yes_____ (2) No_____

       Number of years attended:_____

9. Professional School:

       Graduated: (1) Yes_____ (2) No_____
       Number of year attended:_____

History

10. Place of Birth:_____
                    (State)   (Country)

11. Date of Birth:_____

12. First name of present spouse or significant other:_____

13. First names of past spouses:_____
                                 _____
                                 _____

14. Mother's first name:_____

15. Father's first name:_____

16. First names of brothers and sisters:_____   _____
                                          _____   _____
                                          _____   _____

17. First names of children:_____
                            _____ _____
                            _____   _____

Other Information

18. Ever had epilepsy?   (1) Yes_____   (2) No_____

19. Had recent (within last 5 years) emotional experiences, such as

    Divorce: (1) Yes_____   (2) No_____
    Tragedy in family: (1) Yes_____   (2) No_____
    Other: (1) Yes_____   (2) No_____

20. Ever experienced:

    A vision: (1) Yes_____   (2) No_____ (3) Uncertain_____

21. Ever experienced:

    Contact with a deceased person: (1) Yes_____   (2) No_____ (3) Uncertain_____

22. Ever experienced:

    Telepathy or other "extra sensory perception":
    (1) Yes_____   (2) No_____ (3) Uncertain_____

23. Have you ever felt that someone was controlling your mind by telepathy x-ray, or other unusual means?

    (1) Yes_____   (2) No_____ (3) Uncertain_____

24. Religion.

(1) Catholic_____   (2) Protestant_____   (3) Jewish_____
(4) Other_____   (5) None_____

25. Degree of Involvement in Religion.

(1) No involvement_____   (2) Slight_____   (3) Moderate_____
(4) Deep_____

26. Do you believe in some form of life after the death of the body?

(1) Yes_____   (2) No_____   (3) Uncertain_____

27. Do you believe in reincarnation?

(1) Yes_____   (2) No_____   (3) Uncertain_____

28. ** Are you presently taking any prescription medicines?

(1) Yes_____   (2) No_____

If yes, state the name of the drug:_____

29. Have you consulted with a physician during the past six months?

(1) Yes_____   (2) No_____

30. Do you believe you are in good health, both physically and mentally?

(1) Yes_____   (2) No_____

31. Prior experience with hypnosis?

(1) No experience_____   (2) Prior experience_____

32. If prior experience:

(1) Hypnotic trance was successfully induced_____
(2) Hypnotic trance was attempted but was unsuccessful_____

33. Do you expect to recall past lives under hypnosis?

(1) Yes_____   (2) Probably yes_____   (3) No_____
(4) Probably no_____   (5) Uncertain_____

_____

**In my Second Research Project, the following question was added: Do you believe in the evolution of humans from other species? (1) yes (2) No (3) Uncertain.

APPENDIX B

Statistics-First Research Project

1. General Characteristics of Subjects-First Research Project

Variables	Characteristics	Number of Subjects	Percentage of Total
a. Age	21 - 30	21	20%
	31 - 40	35	34%
	41 - 50	33	32%
	51 - 60	12	12%
	60 +	3	3%
b. Sex	Male	26	25%
	Female	78	75%
c. Years in College	0	21	20%
	1 - 3	44	42%
	4	22	21%
	5 - 6	8	8%
	7 - 9	8	8%
	No information	1	1%
d. College graduate	Yes	40	38.5%
	No	64	61.5%
e. Religion	Christian	47	45%
	Jewish	4	4%
	Other	34	33%
	None	19	18%
f. Religious involvement	None	37	36%
	Slight	30	29%
	Moderate	22	21%
	Deep	14	13%
	No information	1	1%
g. Belief in life after death	Yes	83	80%
	No	1	1%
	Uncertain	20	19%
h. Belief in Reincarnation	Yes	59	57%
	No	2	2%
	Uncertain	43	41%
i. Expect to recall past-lives	Yes	36	35%
	Probably yes	26	25%
	No	17	16%
	Probably no	3	3%

| | | Uncertain | 22 | 21% |

2. Hypnotic Characteristics-First Research Project

Variables	Characteristics	Number of Subjects	Percentage of Total
a. Prior experience with hypnosis	Yes	34	33%
	No	70	67%
b. If prior experience with hypnosis	Trance induced	27	79%
	Trance not induced	7	21%
c. Susceptibility hand test*	No response	8	8%
	Light response	19	18%
	Medium response	31	30%
	Excellent response	45	43%
	Negative response	1	1%
d. Depth Reports	0	1	1%
	2	5	4%
	3	8	7%
	4	6	5%
	5	13	13%
	6	21	20%
	7	21	20%
	8	16	15%
	9	5	5%
	10	4	4%
	30	1	1%
	No report	3	3%

* Before placing each subject into hypnosis, I used a susceptibility test where the subject stands with eyes closed, hands outstretched. I suggest to the subject to visualize that I place a book on one hand which is forcing his/her hand down; and that I tie a helium-filled baloon to the other hand, drawing that hand up.

3. Near-Death Type Characteristics-First Research Project

Variables	Characteristics	Number of Subjects	Percentage of Total
a. Past Life Review	Yes	0	0%
	No	74	71%
	Uncertain	1	1%
	No response	29	28%
b. See light after death	Yes	8	8%
	No	66	63%
	No response	30	29%
c. Others with subject after death	Yes	27	26%
	No	26	25%
	Uncertain	4	4%
	No response	47	45%
d. Special being in other realm	Yes	2	2%
	No	72	69%
	No response	30	29%

4. Regresion Experiences-First Research Project

Variables	Characteristics	Number of Subjects	Percentage of Total
a. Regress beyond birth	Yes	81	78%
	No	23	21%
b. Degree of detail in regression -1st life	Poor	24	30%
	Fair	33	41%
	Good	21	25%
	Excellent	3	4%
c. Degree of detail in regression -2nd life	Poor	14	17%
	Fair	34	42%
	Good	24	30%
	Excellent	2	2%
	No rating	7	9%
d. Regression continent -1st life	North America	22	27%
	South America	1	1%
	Europe	26	32%
	Africa	5	6%
	Near East	5	6%
	Asia	1	1%
	Unknown	2	2%
	No response	19	25%
e. Regression nation -1st life	USA	23	28%
	Canada	1	1%
	England	8	10%
	France	5	6%
	Italy	1	1%
	Egypt	2	2%
	Ireland	3	4%
	Scotland	2	2%
	Germany	1	1%
	Other	13	16%
	Unknown	4	5%
	No response	20	24%
f. Regression continent -2nd life	North America	25	31%
	South America	2	2%
	Europe	21	26%
	Africa	5	6%
	Near East	2	2%
	Asia	2	2%
	Other	2	2%
	No response	22	29%

4. Regression Experiences First Research Project continued

Variables	Characteristics	Number of Subjects	Percentage of Total
g. Regression nation -2nd life	USA	21	26%
	Canada	1	1%
	Mexico	1	1%
	England	3	4%
	France	4	5%
	Russia	2	2%
	Italy	3	4%
	Egypt	3	4%
	China	1	1%
	Ireland	1	1%
	Germany	2	2%
	Other	8	10%
	Unknown	3	4%
	No response	28	35%
h. Time periods subjects regressed to -1st life	1 - 100	2	2%
	101 - 500	1	1%
	501 - 1000	3	4%
	1001 - 1500	2	2%
	1501 - 1600	2	2%
	1601 - 1700	3	4%
	1701 - 1800	7	9%
	1801 - 1850	11	14%
	1851 - 1860	5	6%
	1861 - 1870	5	5%
	1871 - 1880	2	2%
	1881 - 1890	1	1%
	1891 - 1900	3	4%
	1901 - 1950	8	10%
	1951 - 1957	2	2%
	No response	24	31%
i. Time periods subjects regressed to -2nd life	1 - 100	0	0%
	101 - 500	2	2%
	501 - 1000	5	6%
	1001 - 1500	3	4%
	1501 - 1600	2	2%
	1601 - 1700	3	4%
	1701 - 1800	7	9%
	1801 - 1850	8	10%
	1851 - 1860	2	2%
	1861 - 1870	1	1%
	1871 - 1880	3	4%
	1881 - 1890	3	4%
	1891 - 1900	1	1%

4. Regression Experiences First Research Project continued

Variables	Characteristics		Number of Subjects	Percentage of Totals
	1901 - 1950		8	10%
	No response		33	41%
j. Cause of regression death - 1st life	Natural		42	52%
	Diseased	5		6%
	Accidental		10	12%
	Violent		9	11%
	Suicide		1	1%
	Unknown		7	9%
	No response		7	9%
k. Cause of regression death - 2nd life	Natural		27	33%
	Diseased	9		11%
	Accidental		7	9%
	Violent		12	15%
	Other		1	1%
	Unknown		4	5%
	No response		21	26%
l. After death sense of well being -1st life	Yes		35	43%
	No		2	3%
	Don't know		1	1%
	Uncertain		5	6%
	No response		38	47%
m. After death sense of wellbeing -2nd life	Yes		14	17%
	No		3	4%
	Uncertain		1	1%
	No response		63	78%
n. Enlightened in other realm	Yes		10	12%
	No		35	43%
	Uncertain		8	10%
	No response		28	35%
o. Punished in other realm	Yes		0	0%
	No		47	58%
	No response		34	42%
p. Rebirth after past-life death	Chose rebirth		39	48%
	Didn't choose rebirth		23	29%
	Uncertain		4	5%
	Unknown		2	2%
	No response		13	16%

4. Regression Experiences First Research Project continued

Variables	Characteristics	Number of Subjects	Percentage of Totals
q. Help with rebirth choice	Help with choice	14	17%
	No help with choice	25	31%
	Unknown	1	1%
	Uncertain	2	2%
	No response	39	49%
r. See earth-like scenes in other realm	Yes	7	9%
	No	50	62%
	Don't know	1	1%
	Uncertain	7	9%
	No response	16	19%
s. Regress to different sex - 1st life	Yes	15	19%
	No	65	80%
	Uncertain	1	1%
t. Regress to different sex - 2nd life	Yes	25	31%
	No	48	59%
	Uncertain	1	1%
	No response	7	9%
u. Regress to different race - 1st life	Yes	23	28%
	No	53	65%
	Uncertin	2	3%
	No response	3	4%
v. Regress to different race - 2nd life	Yes	23	29%
	No	44	54%
	Uncertain	1	1%
	No response	13	16%
w. Know people from past-life in this life	Yes	21	26%
	No	3	4%
	Uncertain	2	2%
	No resonse	55	68%
x. Knowledge of past-life period	Extensive	2	2%
	Some knowledge	14	17%
	No knowledge	63	79%
	No response	2	2%
y. Opinion of past-life experience	Imagined	8	10%
	Real	42	52%
	Uncertain	25	31%
	No response	6	7%

4. Regression Experiences First Research Project continued.

Variables	Characteristics	Number of Subjects	Percentage of Totals
z. Has regression About life			
changed feelings		2	2%
Religious concepts		3	4%
Fear of death		2	2%
Other		3	4%
None		26	32%
Uncertain		8	10%
No response		37	46%
ab. Lived on another	Yes	30	37%
planet	No	31	38%
	Unknown	8	10%
	Uncertain	7	9%
	No resonse	5	6%

5. Relationships between selected Variables-First Research Project.

a. Religious Beliefs, Religious Involvement, Expectations, Education and Regression to Past-Lives.

The data generated by my First Research Project indicate that a person's religious beliefs, religious involvement, belief in life after death, education and whether he/she expected to recall past-lives had no effect on whether or not he/she regressed beyond birth while in hynosis, and made contact with a past-life.

(1). Religion compared with Regress beyond Birth produced no statistical significance. p >.05. (Chi Square .88411).*

(2). Religious Involvement compared with Regress beyond Birth produced no statistical significance. p >.05 (Chi-Square .36561).

(3). Belief in Life After Death compared with Regress beyond Birth produced no statistical significance. p >.05 (Chi-Square .56712).

---

* "Statistical Significance" generally refers to the degree to which an obtained value will not occur by chance and can therefore be attributed to another factor. A convention in general use has established the .05 level as the minimum significance level, indicating that the observed difference is real; such results would occur 5 percent of the time by chance.

(4). Belief in Reincarnation compared with Regress beyond Birth produced no statictical significance. p >.05 (Chi-Square .70747).

(5). Education compared with Regress beyond Birth produced no statistical significance. p >.05 (Chi-Square .12923).

(6). Expect to Recall Past-Lives compared with Regress beyond Birth produced no statistical significnce. p >.05 (Chi-Square .82524).

b. Experience with Hypnosis.

The data from my First Research Project indicate the Depth in the Hypnotic Trance that a person achieved did have an effect on whether the person Regressed Beyond Birth, and also had an effect on the amount of Detail Given in the Regression,.
(1). Depth of Trance compared with Regress beyond Birth produced a statistical significance. p <.05. (Chi-Square .03711).

Further analysis, using Analysis of Variance, revealed that the Depth of Trance achieved by those persons who did not Regress Beyond Birth,was significantly different when compared to the Depth of Trance of those who did Regress Beyond Birth. (F (1,99)=8.30, p < .01).

Persons who did not Regress Beyond Birth reported a mean of 4.68 for the Depth of Trance response. Persons who did Regress Beyond Birth reported a mean of 6.79. On a depth scale of 1 to 10, these results suggest that a person with a depth score of 5 or lower, would be less likely to Regress Beyond Birth, than a person who has a depth score of 7 or above.
(2). Depth of Trance compared with Degree of Detail in Regression produced a statistical significance. p <.05 (Chi-Square .03439).

Further analysis, using a Person Product-Moment Correlation, disclosed a positive relationship between the Depth of Trance reported and the Degree of Detail given in the Regression Beyond Birth. (r = .34 ** p < .01). In other words, as the Depth of Trance increases, the Degree of Detail in the hypnotic regression increases.

APPENDIX C
Statistics-Second Research Project

1. General Characteristics of Subjects-Second Research Project

Variables	Characteristics	Number of Subjects	Percentage of Total
a. Age	18	1	2%
	21 - 30	6	12%
	31 - 40	9	18%
	41 - 50	21	42%
	51 - 60	9	18%
	61+	4	8%
b. Sex	Male	7	14%
	Female	43	86%
c. Race	White	47	94%
	Black	2	4%
	Other	1	2%
d. Religion	Christian	18	36%
	Jewish	3	6%
	Other	21	42%
	None	8	16%
e. Religious Involvement	None	21	42%
	Slight	11	22%
	Moderate	9	18%
	Deep	7	14%
	No Information	2	4%
f. Belief in Life after Death	Yes	44	88%
	No	0	0%
	Uncertain	6	12%
g. Belief in Reincarnation	Yes	37	74%
	No	0	0%
	Uncertain	12	24%
	No information	1	2%
h. Belief in Evolution	Yes	19	38%
	No	13	26%
	Uncertain	17	34%
	No information	1	2%
i. Graduate from High School	Yes	49	98%
	No	1	2%

1. General Characteristics of Subjects-Second Research Project continued

Variables	Characteristics	Number of subjects	Percentage of Total
j. Years in College	0 - 2	31	62%
	3 - 4	10	20%
	5+	9	18%
k. Expect to Recall Past Lives	Yes	17	34%
	Probably Yes	14	28%
	No	0	0%
	Probably No	1	2%
	Uncertain	18	36%
l. Susceptibility Hand Test	No Response	4	8%
	Light Response	12	24%
	Medium Response	9	18%
	Excellent Response	25	50%
m. Regress to Past Lives	Yes	44	88%
	No	6	12%
n. Depth Report	0 - 4	9	18%
	5 - 10	40	80%
	No information	1	2%
o. Degree of Detail in Regression	Poor	9	20%
	Fair	16	36%
	Good	16	36%
	Excellent	3	8%
p. Verifiable Data	Yes	19	43%
	No	25	56%
q. Regress to Different Sex	Yes	24	55%
	No	20	45%
r. Regress to Different Race	Yes	26	59%
	No	17	39%
	No information	1	2%

1. General Characteristics of Subjects-Second Research Project continued

Variables	Characteristics	Number of Subjects	Percentage of Total
s. Location after Death	In the Air	32	73%
	Unknown	2	5%
	Can't Describe	1	2%
	Other	5	11%
	No information	4	9%
t. Others With Personality after Death	Yes	20	45%
	No	24	55%
u. Know People from Past Life in this Life	Yes	19	44%
	No	12	27%
	Uncertain	1	2%
	No information	12	27%
v. Time Spirit enters Fetus	At Conception	1	2%
	Fetus in Human Form	25	57%
	Fetus not in Human Form	10	23%
	At Birth	1	2%
	No information	7	16%
w. Aware of Fetus Development	Yes	13	30%
	No	4	9%
	No information	27	61%
x. Aware of Mother's Feelings while a Fetus	Yes	30	68%
	No	4	9%
	No information	10	23%
y. First Life on Earth	Human	30	68%
	Subhuman	5	11%
	Animal	2	5%
	Spirit	2	5%
	Other	1	2%
	No information	4	9%
z. Lived on Another Planet	Yes	13	30%
	No	20	46%
	Uncertain	5	11%
	Unknown	1	2%
	No information	5	11%

2. Relationships between selected Variables-Second Research Project

a. Religious Beliefs, Religious Involvement, Expectations, Education, Evolution, and Regression to Past-Lives.

    The data generated by my Second Research Project indicate that a person's religious beliefs, religious involvement, belief in life after death, education, belief in evolution, and whether he/she expected to recall-past lives had no effect on whether or not he/she regressed beyond birth while in hypnosis and made contact with a past-life.

    (1). Religion compared with Regress Beyond Birth produced no statistical significance. $;>.05$. (Chi-Square .37610).[*]

    (2). Religious Involvement compared with Regress Beyond Birth produced no statistical significance. $p>.05$. (Chi-Square .32949).

    (3). Belief in Life After Death compared with Regress Beyond Birth produced no statistical significance. $p>.05$. (Chi-Square .08649).

    (4). Belief in Reincarnation compared with Regress Beyond Birth produced no statistical significance. $p>.05$. (Chi-Square .59076).

    (5). Education compared with Regress Beyond Birth produced no statistical significance. $p>,05$. (Chi-Square .75229).

    (6). Expect to Recall Past-Lives compared with Regress Beyond Birth produced no statistical significance. $p>.05$. (Chi-Square .07156).

    (7). Belief in Evolution compared with Regress Beyond Birth produced no statistical significance. $p>.05$. (Chi-Square .78799).

b. Experience with Hypnosis.

    The data from my Second Research Project indicate that the depth in the hypnotic trance that a person achieved did have an effect on whether the person regressed beyond birth. It also had an effect on the amount of detail given in the regression, and the data given in the regression, if true, that can be verified

    (1). Depth of Trance compared with Regress Beyond Birth produced a very strong statistical significance. $p<.05$ (Chi-Square .00000).

    (2). Depth of Trance compared with Degree of Detail in Regression produced a strong statistical significance. $p<.05$. (Chi-Square .00456).

    (3). Depth of Trance compared with Verifiable Data produced a strong statistical significance. $p<.05$ (Chi-Square .00823).

---

[*] "Statistical Significance" generally refers to the degree to which an obtained value will not occur by chance and can therefore be attributed to another factor. A convention in general use has established the .05 level as the minimum significance level, indicating that the observed difference is real; such results would occur 5 percent of the time by chance.

APPENDIX D

Verifiable Past Lives?

The names and other information listed below identify the personalities to whom several of the participants in my Research Projects regressed. The names are spelled phonetically. The dates may or may not be accurate as in some cases the dates of birth and the dates the participants gave when reaching later ages are not consistent.

1. **Jonathon Morse**. A white male, Jonathon lived on a farm near Atlanta, Georgia, in the early 1860s. He lived in a large, stone house with pillars, and with a big, white front door. Inside the house was a large stairway going up to the second floor. The family had servants.

Jonathon's mother was named Cynthia and he called his father Colonel David. Lincoln was President. Jonathon died in the Civil War at age nineteen, a member of the Confederate Army.

2. **Diana Anderson**. A white female with reddish hair, wearing glasses. At age twenty-seven she lived at 2347 Otis Avenue or street, in Evansville, Ohio, with her husband Jim Anderson. They had no children. She had neighbors, "The Sheppards". She stated the year was 1910. Her father was named Bill Evans and her mother was Mora Evans and they lived in Philadelphia. At age thirty-five she lived in the same house. Her parents were both dead at that time. She was born in Illinois and died at age fifty-seven. An Internet search shows an Evansville, Ohio on Salt Springs Road, highway No. 64, Approximately 1½ miles West of McDonald, Ohio.

3. **Sarah Baker**. A white woman, born in New York. Her father was George Ashley and her mother Dorothea. The 1820 US Census shows a George Ashley living in New York City, in the Eighth ward. Sarah married Joseph Baker, and lived in Allen Pennsylvania. At the age of 30, the year was possibly about 1843.

Sarah had two children George and Sarah, and was pregnant with a third when she was about thirty-years old. The 1850 US Census shows a Joseph Baker, a farmer, age forty-three. His wife Sarah was thirty-six. They had four children May, Francis, George and Elizabeth. Sarah died at age thirty-eight of a flu type disease.

4. **Lola Sanchez**. A white female. At age twenty she was living in Charleston, South Carolina. The year is 1920. Her mother Rita Sanchez and her father Frank Sanchez are both deceased. She died at age thirty-five in a car accident.

5. **Ronald Ayers**. Ronald is a white male, age forty-seven, an inmate in a prison in England. The year is 1792. Ronald had a mate prior to this time named Elizabeth. His father was Arnold and his mother Marina.

Ronald killed a man with a club who had tried to steal from him. Ronald was hung for his crime at age forty-seven.

6. **Sarah Aletha Hill.** A white woman, age twenty-three. She ran "The Manor" which is near San Francisco, California. The year is 1857. She is present at the dedication of an orphanage. The Governor is also there.

At twenty-three, her parents are both deceased. Her father was Samuel Hill; her mother Elizabeth hill. She is single and has no children.

At age thirty-five she has been committed to an asylum in San Francisco. She says the reason she was committed was "…because I found out they were going to kill President Lincoln." At thirty-five, the year would probably have been 1869. She died at the asylum at age eighty-five. This would have been about 1921.

7. **Leanne Neilson.** A female. At age twenty-five she lived in Philadelphia, Pa., married to Paul Neilson who was a Captain in the army. They had two children, John and Michael Neilson. At age sixteen her maiden name was McAlister. She was called Lenny. She lived with her step-mother Nelda and father Milton McAlister in an "old, big house". She had a sister Molly. Molly eventually married three times. At age sixteen, the year was 1784. Leanne died at age forty-two.

8. **Marissa Martin.** A young girl, about the age of nine. The year is about 1910. She lives with her mother Alice Martin and father Bruce Martin, in Winnetka, Illinois. Her father has a store and sells tobacco. She has a brother Tom and a sister. She died at about the age of nine or ten.

9. **Konla Jameson.** A black male, forty-years of age. He has a mate Tomeka and a daughter Kara. They live in Atlanta, George. The year is 1895. They attended the wedding of their daughter Kara in a "new Baptist Church." His father's name was Kenneth Jameson and his mother's was Meredith Jameson. Konla died at age ninety, and his wife survived him.

10. **Ronnie Copeland.** A seventeen-year-old male, living in Kansas City, Missouri. His father was Howard Copeland, who built houses, and his mother is Vicky Copeland. The year is 1963. Ronnie is killed in an automobile accident "playing chicken" at the age of seventeen.

11. **Andra Pickens.** Andra is a white single female, age twenty-five-years. She is living in an Indian Village in Oklahoma, giving medical aid to members of an Indian tribe. The year is not known. Her father's name is Bill Pickens and her mother's is Susan Pickens. Susan is probably deceased. The father has tuberculosis and gave the disease to Andra.

12. **Ethel Smith.** A twenty-five-year-old white woman married to Ralph Smith, living on a farm near Lincoln, Kansas. The year is about 1870. Her father is Robert Johnson and her mother is Ethel Johnson, and she has a twin sister named Inez. In the 1870s, she attended a graduation ceremony at McPherson,

Kansas where her cousin Bill was graduating. She was born in Indiana. At age seventeen she lived with her parents on a farm in Indiana. She died at the age twenty-six in 1872.

13. **Scott Harris**. A black male living in New York City at the age of forty. He has a wife, Kathy Harris, and three children, Karen, Calvin, and one other. The year is 1942. Scott works in an office where he wears a suit to work. He is Catholic. He died in a hospital at the age of fifty-five of a heart attack. Scott was born in Georgia, near Atlanta. His father was Andrew Harris and his mother was Sarah Harris. His father worked as a miner. When he was ten years old, a white girl named Korin lived with them.

14. **Eleanor Pritchard**. At age twenty-two, Eleanor is married to John Pritchard and they lived on a farm in Kansas. The nearest town is Wichita but it is quite a distance away. They have two children, Amy and Tyler. The year is 1868. At age thirty, Eleanor is living on the farm alone and she believes John is dead. He left five years ago and didn't return. Eleanor died at the age of 38 of influenza. There was a physician in attendance named Timothy O'Leary.

15. **Annie Miller**. At age thirty Annie, a white female, lives with her husband John Miller in (or near) Princeton, Colorado The time is in the 1950s. They have no children "as yet." John is in the construction business. At age ten Annie lived in Atlanta, Georgia with her father Jim Hornesby, a welder by trade, and Anna Hornesby, her mother. In the year 1950 she still lived in Atlanta and was not married, although she and John were lovers. Annie died in St. Elmo, Colorado, as a result of injuries she incurred in a fire. St. Elmo is now almost a ghost town located near Mount Princeton. An Internet search shows a town named Princeton that once existed in central Colorado on highway No. 24, North of Buena Vista between highway markers 371 and 388, just East of some railroad tracks.

16. **Elizabeth Bursen**. A white single female, age thirty-nine, living in Vermont. She has no mate and no children. She states that the year is 1914. Her mother, Mariam Bursen, is alive, but her father, Stanford Bursen, is deceased. She was born in Pineville, Oregon in 1806. (The dates don't match up of course.) She stated Pineville is "upstate." She died at the age of fifty-four in a town she described as Bergmount, Vermont. Pineville, Oregon is located on US Highway No. 26 at the junction with State Highway 126. I have not found a Bergmount, Vermont to date.

17. **Lilius Hansen**. A white female, in her thirties married to Howard Hansen. They have no children. The year is 1875 and they lived in Vermont. The nearest town is Springdale. Her parents, Gladys Ormesby and Henry Ormesby.(Phonetic: might be Ornesby, Ormesbee, or something similar.) Both

of her parents are deceased. She was born in New Hampshire. My search to date has not found a town named *Springdale* in Vermont.

18. **Suzy Brown**. From ages ten to twenty-five Suzy lived with Margaret Brown and Edward Brown in Hartford, Connecticut. At age fifteen, the year was 1895. Suzy can't read or write. After Margaret's death, Suzy went west to Elkton, Nevada where she worked as a cook. At age forty she was married to John Jewmus. John worked on the railroad. Suzy died at age forty-three from falling down a flight of stairs. After the session, the subject told me that Suzy was retarded. I have found an Elkhorn in Nevada, but not an Elkton.

19. **Brenda Williams**. A white female at age twenty-two married to Peter Williams. Brenda and Peter had one child, a son Peter. The year is 1928. They lived at 212 Grove Street, Philadelphia, Pennsylvania. Brenda's mother Rebecca Johnson and father William Johnson were both deceased. At age thirty Brenda lived in the same house and had a cook named Mrs. James. Brenda was born in Chillicothe, Ohio and had no brothers or sisters. She died in an automobile accident at an unknown date.

20. **Mattie Dillinger**. A white female probably a widow at age twenty-six. She has one daughter Ruth Ann Dillinger. Her mate was Paul and was killed "in a war." She lives in the country near Waco, Texas. The year is 1926. Her father was Charles Hayes and her mother was Ruth Hayes. Both of her parents were alive in 1926. Mattie died at the age of seventy in a nursing home.

21. **Matthew Brown**. Matthew is male and was born in 1932 in St. Louis, Missouri. He stated that he had brown skin. His father was Hank Brown and his mother was June Brown. The year is 1950 and Matthew is still in St. Louis. Both of his parents are deceased. Matthew died at the age of forty-two of a heart attack.

22. **Rainbow McKenzie**. A white female. At age twenty-five she was living with friends in the State of New York. The city may be New York City. The year is 1966. Rainbow's parents Robert McKenzie and Robin McKenzie were both deceased in 1966. Rainbow was born in New York in 1941. She died of a drug overdose at the age of twenty-nine.

If you, the reader, have any information about any of these people, or if you live in the area where any of these people lived and would like to volunteer to do a little research, by all means contact me as follows:

Robert T. James
1983 North Academy
Colorado Springs, Colorado 80909-1503
(719) 597-8803.
Email:rjames66@Juno.com
Web site: www.hypnoti.st

About the Author

Robert T. James lives in Colorado Springs, Colorado. He received a Bachelor of Science Degree as well as a Juris Doctor Degree from the University of Denver. Dr. James practiced law for thirty-five years and, after retiring in 1986, studied psychology and psychotherapy for three years at the University of Colorado in Colorado Springs.

He has been a student of hypnosis for over forty years, attending numerous schools and workshops furthering his knowledge of hypnosis, hypnotherapy and forensic hypnosis. The American Council of Hypnotist Examiners has certified Dr. James as both a Master Hypnotist and a Hypnotherapist.

He has authored numerous published articles on the past-life phenomena.

His web site is at www.hypnoti.st

0-595-31022-2

Printed in the United States
92608LV00004B/189/A